REMAKING
E D E N

Cloning and Beyond in
a Brave New World

Lee M. Silver

WEIDENFELD & NICOLSON
LONDON

First published in Great Britain in 1998
by Weidenfeld & Nicolson

This edition published by arrangement with Avon Books,
A division of The Hearst Corporation,
1350 Avenue of the Americas, New York NY 10019

A CIP catalogue record for this book is available
from the British Library.

ISBN 0 297 84135 1

Printed in Great Britain by Clays Ltd, St Ives plc

Weidenfeld & Nicolson
The Orion Publishing Group Ltd
Orion House
5 Upper Saint Martin's Lane
London, WC2H 9EA

To Joseph and Ethel Silver
for creating me the old-fashioned way

■

To Sylvia Goodman, the memory of Leon Goodman,
and Joseph and Rebecca Silver
for genes and memes

■

Et comme toujours,
à la femme de ma vie et mes enfants

Contents

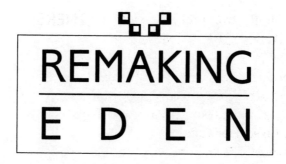

O, wonder!
How many goodly creatures are there here!
How beauteous mankind is!
O brave new world
That has such people in't

—William Shakespeare, *The Tempest*

Prologue
A Glimpse of Things to Come

Sometime in the not-so-distant future, you may visit the maternity ward at a major university hospital to see the newborn child or grandchild of a close friend. The new mother, let's call her Barbara, seems very much at peace with the world, sitting in a chair quietly nursing her baby, Max. Her labor was—in the parlance of her doctor—"uneventful," and she is looking forward to raising her first child. You decide to make pleasant conversation by asking Barbara whether she knew in advance that her baby was going to be a boy. In your mind, it seems like a perfectly reasonable question since doctors have long given prospective parents the option of learning the sex of their child-to-be many months before the predicted date of birth. But Barbara seems taken aback by the question. "Of course I knew that Max would be a boy," she tells you. "My husband Dan and I chose him from our embryo pool. And when I'm ready to go through this again, I'll choose a girl to be my second child. An older son and a younger daughter—a perfect family."

Now, it's your turn to be taken aback. "You made a conscious choice to have a boy rather than a girl?" you ask.

"Absolutely!" Barbara answers. "And while I was at it, I made sure that Max wouldn't turn out to be fat like my brother Tom or addicted to alcohol like Dan's sister Karen. It's not that I'm personally biased or anything," Barbara continues defensively. "I just wanted to make sure that Max would have the greatest chance for achieving success. Being overweight or alcoholic would clearly be a handicap."

You look down in wonderment at the little baby boy destined to be moderate in both size and drinking habits.

Max has fallen asleep in Barbara's arms, and she places him gently in his bassinet. He wears a contented smile, which evokes a similar smile from his mother. Barbara feels the urge to stretch her legs and asks whether you'd like to meet some of the new friends she's made during her brief stay at the hospital. You nod, and the two of you walk into the room next door where a thirty-five-year old woman named Cheryl is resting after giving birth to a nine-pound baby girl named Rebecca.

Barbara introduces you to Cheryl as well as a second woman named Madelaine, who stands by the bed holding Cheryl's hand. Little Rebecca is lying under the gaze of both Cheryl and Madelaine. "She really does look like both of her mothers, doesn't she?" Barbara asks you.

Now you're really confused. You glance at Barbara and whisper, "Both mothers?"

Barbara takes you aside to explain. "Yes. You see Cheryl and Madelaine have been living together for eight years. They got married in Hawaii soon after it became legal there, and like most married couples, they wanted to bring a child into the world with a combination of both of their bloodlines. With the reproductive technologies available today, they were able to fulfill their dreams."

You look across the room at the happy little nuclear family—Cheryl, Madelaine, and baby Rebecca—and wonder how the hospital plans to fill out the birth certificate.

DATELINE SEATTLE: MARCH 15, 2050

You are now forty years older and much wiser to the ways of the modern world. Once again, you journey forth to the maternity ward. This time, it's your own granddaughter Melissa who is in labor. Melissa is determined to experience natural childbirth and has refused all offers of anesthetics or painkillers. But she needs something to lift her spirits so that she can continue on through the waves of pain. "Let me see her pictures again," she implores her husband Curtis as the latest contraction sweeps through her body. Curtis picks the photo album off the table and opens it to face his wife. She looks up at the computer-generated picture of a five-year-old girl with wavy brown hair, hazel eyes, and a round face. Curtis turns the page, and Melissa gazes at an older version of the same child: a smiling sixteen-year-old who is 5 feet, 5 inches tall with a pretty face. Melissa smiles back at the future picture of her yet-to-be-born child and braces for another contraction.

There is something unseen in the picture of their child-to-be that provides even greater comfort to Melissa and Curtis. It is the submicroscopic piece of DNA—an extra gene—that will be present in every cell of her body. This special gene will provide her with lifelong resistance to infection by the virus that causes AIDS, a virus that has evolved to be ever more virulent since its explosion across the landscape of humanity seventy years earlier. After years of research by thousands of scientists, no cure for the awful disease has been found, and the only absolute protection comes from the insertion of a resistance gene into the single-cell embryo within twenty-four hours after conception. Ensconced in its chromosomal home, the AIDS resistance gene will be copied over and over again into every one of the trillions of cells that make up the human body, each of which will have its own personal barrier to infection by the AIDS-causing virus HIV. Melissa and Curtis feel lucky indeed to have the financial wherewithal needed to endow all of their children with this protective agent. Other, less well-off American families cannot afford this luxury.

Outside Melissa's room, Jennifer, another expectant mother, is anxiously pacing the hall. She has just arrived at the hospital and her contractions are still far apart. But, unlike Melissa, Jennifer has no need for a computer printout to show her what her child-to-be will look like as a young girl or teenager. She already has thousands of pictures that show her future daughter's likeness, and they're all real, not virtual. For the fetus inside Jennifer is her identical twin sister—her clone—who will be born thirty-six years after she and Jennifer were both conceived within the same single-cell embryo. As Jennifer's daughter grows up, she will constantly behold a glimpse of the future simply by looking at her mother's photo album and her mother.

DATELINE U.S.A.: MAY 15, 2350

It is now three hundred years later and although you are long since gone, a number of your great-great-great-great-great-great-great-great-great-great-grandchildren are now alive, mostly unbeknownst to one another. The United States of America still exists, but it is a different place from the one familiar to you. The most striking difference is that the extreme polarization of society that began during the 1980s has now reached its logical conclusion, with all people belonging to one of two classes. The people of one class are referred to as *Naturals,* while those in the second class are called the *Gene-enriched* or simply the *GenRich.*

These new classes of society cut across what used to be traditional racial and ethnic lines. In fact, so much mixing has occurred during the last three hundred years that sharp divisions according to race—black versus white versus Asian—no longer exist. Instead, the American populace has finally become the racial melting pot that earlier leaders had long hoped for. The skin color of Americans comes in all shades from African brown to Scandinavian pink, and traditional Asian facial features are present to a greater or lesser extent in a large percentage of Americans as well.

But while racial differences have mostly disappeared, another difference has emerged that is sharp and easily defined. It is the difference between those who are genetically enhanced and those who are not. The GenRich—who account for 10 percent of the American population—all carry synthetic genes. Genes that were created in the laboratory and did not exist within the human species until twenty-first century reproductive geneticists began to put them there. The GenRich are a modern-day hereditary class of genetic aristocrats.

Some of the synthetic genes carried by present-day members of the GenRich class were already carried by their parents. These genes were transmitted to today's GenRich the old-fashioned way, from parent to child through sperm or egg. But other synthetic genes are new to the present generation. These were placed into GenRich embryos through the application of genetic engineering techniques shortly after conception.

The GenRich class is anything but homogeneous. There are many types of GenRich families, and many subtypes within each type. For example, there are GenRich athletes who can trace their descent back to professional sports players from the twenty-first century. One subtype of GenRich athlete is the GenRich football player, and a sub-subtype is the GenRich running back. Embryo selection techniques have been used to make sure that a GenRich running back has received all of the natural genes that made his unenhanced foundation ancestor excel at the position. But in addition, at each generation beyond the foundation ancestor, sophisticated genetic enhancements have accumulated so that the modern-day GenRich running back can perform in a way not conceivable for any unenhanced Natural. Of course, all professional baseball, football, and basketball players are special GenRich subtypes. After three hundred years of selection and enhancement, these GenRich individuals all have athletic skills that are clearly "nonhuman" in the traditional sense. It would be impossible for any Natural to compete.

Another GenRich type is the GenRich scientist. Many of the synthetic genes carried by the GenRich scientist are the same as those carried by all other members of the GenRich class, including some that enhance a variety of physical and mental attributes, as well as others that provide resistance to all known forms of human disease. But in addition, the present-day GenRich scientist has accumulated a set of particular synthetic genes that work together with his "natural" heritage to produce an enhanced scientific mind. Although the GenRich scientist may appear to be different from the GenRich athlete, both GenRich types have evolved by a similar process. The foundation ancestor for the modern GenRich scientist was a bright twenty-first-century scientist whose children were the first to be selected and enhanced to increase their chances of becoming even brighter scientists who could produce even more brilliant children. There are numerous other GenRich types including GenRich businessmen, GenRich musicians, GenRich artists, and even GenRich intellectual generalists who all evolved in the same way.

Not all present-day GenRich individuals can trace their foundation ancestors back to the twenty-first century, when genetic enhancement was first perfected. During the twenty-second and even the twenty-third centuries, some Natural families garnered the financial wherewithal required to place their children into the GenRich class. But with the passage of time, the genetic distance between Naturals and the GenRich has become greater and greater, and now there is little movement up from the Natural to GenRich class. It seems fair to say that society is on the verge of reaching the final point of complete polarization.

All aspects of the economy, the media, the entertainment industry, and the knowledge industry are controlled by members of the GenRich class. GenRich parents can afford to send their children to private schools rich in the resources required for them to take advantage of their enhanced genetic potential. In contrast, Naturals work as low-paid service providers or as laborers, and their children go to public schools. But twenty-fourth-century public schools have little in common with their predecessors from the twentieth century. Funds for public education have declined steadily since the beginning of the twenty-first century, and now Natural children are only taught the basic skills they need to perform the kinds of tasks they'll encounter in the jobs available to members of their class.

There is still some intermarriage as well as sexual intermingling between a few GenRich individuals and Naturals. But, as one might imagine, GenRich parents put intense pressure on their children not to dilute their expensive genetic endowment in this way. And as time passes, the mixing of the classes will become less and less frequent for reasons of both environment and genetics.

The environmental reason is clear enough: GenRich and Natural children grow up and live in segregated social worlds where there is little chance for contact between them. The genetic reason, however, was unanticipated.

It is obvious to everyone that with each generation of enhancement, the genetic distance separating the GenRich and Naturals is growing larger and larger. But a startling consequence of the expanding genetic distance has just come to light. In a nationwide survey of the few interclass Gen-Rich-Natural couples that could be identified, sociologists have discovered an astounding 90 percent level of infertility. Reproductive geneticists have examined these couples and come to the conclusion that the infertility is caused primarily by an incompatibility between the genetic makeup of each member.

Evolutionary biologists have long observed instances in which otherwise fertile individuals taken from two separate populations prove infertile when mated to each other. And they tell the sociologists and the reproduc-

tive geneticists what is going on: the process of species separation between
the GenRich and Naturals has already begun. Together, the sociologists,
the reproductive geneticists, and the evolutionary biologists are willing to
make the following prediction: If the accumulation of genetic knowledge
and advances in genetic enhancement technology continue at the present
rate, then by the end of the third millennium, the GenRich class and the
Natural class will become the GenRich humans and the Natural humans—
entirely separate species with no ability to cross-breed, and with as much
romantic interest in each other as a current human would have for a
chimpanzee.

DATELINE PRINCETON, NEW JERSEY: THE PRESENT

Are these outrageous scenarios the stuff of science fiction? Did they spring
from the minds of Hollywood screenwriters hoping to create blockbuster
movies without regard to real world constraints? No. The scenarios described under the first two datelines emerge directly from scientific understanding and technologies that are already available today. The scientific
framework for the last scenario is based on straightforward extrapolations
from our current knowledge base. Furthermore, if biomedical advances
continue to occur at the same rate as they do now, the practices described
are likely to be feasible long before we reach my conservatively chosen
datelines.

It's time to take stock of the current state of science and technology
in the fields of reproduction and genetics and to ask, in the broadest terms
possible, what the future may hold. Most people are aware of the impact
that reproductive technology has already had in the area of fertility treatment. The first "test tube baby"—Louise Brown—is already eighteen years
old, and the acronym for in vitro fertilization—IVF—is commonly used
by laypeople. The cloning of human beings has become a real possibility
as well, although many are still confused about what the technology can
and cannot do. Advances in genetic research have also been in the lime-
light, with the almost weekly identification of new genes implicated in
diseases like cystic fibrosis and breast cancer, or personality traits like
novelty-seeking and anxiety.

What has yet to catch the attention of the public at large, however, is the incredible power that emerges when current technologies in reproductive biology and genetics are brought together in the form of *reprogenetics*. With reprogenetics, parents can gain complete control over their genetic destiny, with the ability to guide and enhance the characteristics of their children, and their children's children as well. But even as reprogenetics makes dreams come true, like all of the most powerful technologies invented by humankind, it may also generate nightmares of a kind not previously imagined.

Of course, just because a technology becomes feasible does not mean that it will be used. Or does it? Society, acting through government intervention, could outlaw any one or all of the reprogenetic practices that I have described. Isn't the *nonuse* of nuclear weapons for the purpose of mass destruction over the last half century an example of how governments can control technology?

There are two big differences between the use of nuclear technology and reprogenetic technology. These differences lie in the resources and money needed to practice each. The most crucial resources required to build a nuclear weapon—large reactors and enriched sources of uranium or plutonium—are tightly controlled by the government itself. The resources required to practice reprogenetics—precision medical tools, small laboratory equipment, and simple chemicals—are all available for sale, without restriction, to anyone with the money to pay for them. The cost of developing a nuclear weapon is billions of dollars. In contrast, a reprogenetics clinic could easily be run on the scale of a small business anywhere in the world. Thus, even if restrictions on the use of reprogenetics are imposed in one country or another, those intent on delivering and receiving these services will not be restrained. But on what grounds can we argue that they should be restrained?

In response to this question, many people point to the chilling novel *Brave New World* written by Aldous Huxley in 1931. It is the story of a future worldwide political state that exerts complete control over human reproduction and human nature as well. In this brave new world, the state uses fetal hatcheries to breed each child into a predetermined intellectual class that ranges from alpha at the top to epsilon at the bottom. Individual members of each class are predestined to fit into specific roles in a soulless utopia where marriage and parenthood are prevented and promiscuous sexual activity is strongly encouraged, where universal immunity to diseases has been achieved, and where an all-enveloping state propaganda machine and mood-altering drugs make all content with their positions in life.

While Huxley guessed right about the power we would gain over the process of reproduction, I think he was dead wrong when it came to predicting *who* would use the power and for what purposes. What Huxley failed to understand, or refused to accept, was the driving force behind babymaking. It is individuals and couples who want to reproduce themselves in their own images. It is individuals and couples who want their children to be happy and successful. And it is individuals and couples—like Barbara and Dan and Cheryl and Madelaine and Melissa and Curtis and Jennifer, *not governments*—who will seize control of these new technologies. They will use some to reach otherwise unattainable reproductive goals and others to help their children achieve health, happiness, and success. And it is in pursuit of this last goal that the combined actions of many individuals, operating over many generations, could perhaps give rise to a polarized humanity more horrific than Huxley's imagined Brave New World.

There are those who will argue that parents don't have the right to control the characteristics of their children-to-be in the way I describe. But American society, in particular, accepts the rights of parents to control every other aspect of their children's lives from the time they are born until they reach adulthood. If one accepts the parental prerogative after birth, it is hard to argue against it before birth, if no harm is caused to the children who emerge.

Many think that it is inherently unfair for some people to have access to technologies that can provide advantages while others, less well-off, are forced to depend on chance alone. I would agree. It is inherently unfair. But once again, American society adheres to the principle that personal liberty and personal fortune are the primary determinants of what individuals are allowed and able to do. Anyone who accepts the right of affluent parents to provide their children with an expensive private school education cannot use "unfairness" as a reason for rejecting the use of reprogenetic technologies.

Indeed, in a society that values individual freedom above all else, it is hard to find any legitimate basis for restricting the use of reprogenetics. And therein lies the dilemma. For while each individual use of the technology can be viewed in the light of personal reproductive choice—with no ability to change society at large—together they could have dramatic, unintended, long-term consequences.

As the technologies of reproduction and genetics have become ever more powerful over the last decade, most practicing scientists and physicians have been loathe to speculate about where it may all lead. One reason for reluctance is the fear of getting it wrong. It really is impossible to predict with certainty which future technological advances will proceed on time and which will encounter unexpected roadblocks. This means that like Huxley's vision of a fetal hatchery, some of the ideas proposed here may ultimately be technically impossible or exceedingly difficult to implement. On the other hand, there are sure to be technological breakthroughs that no one can imagine now, just as Huxley was unable to imagine genetic engineering, or cloning from adult cells, in 1931.

There is a second reason why fertility specialists, in particular, are reluctant to speculate about the kinds of future scenarios that I describe here. It's called politics. In a climate where abortion clinics are on the alert for terrorist attacks, and where the religious right rails against any interference with the "natural process" of conception, IVF providers see no reason to call attention to themselves through descriptions of reproductive and genetic manipulations that are sure to provoke outrage.

The British journal *Nature* is one of the two most important science journals in the world (the other being the American journal *Science*). It is published weekly and is read by all types of scientists from biologists to physicists to medical researchers. No one would ever consider it to be radical or sensationalist in any way. On March 7, 1996, *Nature* published an article that described a method for cloning unlimited numbers of sheep from a single fertilized egg, with further implications for improving methods of genetic engineering. It took another week before the ramifications of this isolated breakthrough sank in for the editors. On March 14, 1996, they wrote an impassioned editorial saying in part: "That the growing power of molecular genetics confronts us with future prospects of being able to *change the nature of our species* [my emphasis] is a fact that seldom appears to be addressed in depth. Scientific knowledge may not yet permit detailed understanding, but the possibilities are clear enough. This gives rise to issues that in the end will have to be related to people within the social and ethical environments in which they live. . . . And the agenda is set by mankind as a whole, not by the subset involved in the science."

They are right that the agenda will not be set by scientists. But they are wrong to think that "mankind as a whole"—unable to reach consensus on so many other societal issues—will have any effect whatsoever. The agenda is sure to be set by individuals and couples who will act on behalf of themselves and their children.

In the pages that follow, I will explain how remarkable advances in science and technology force us to reconsider long-held notions of parenthood, childhood, and the meaning of life itself. I will show you how technological advances, in particular, provide individuals and couples of all kinds with options for reproducing in ways that were previously unimaginable. And I will present multiple perspectives on a future in which people use reprogenetics to assume control over the destiny of humankind.

Throughout, I will explore the many objections that have been raised to the use of this technology. In some instances, I will attribute opposition to conscious or subconscious fears of treading in "God's domain." But in all cases, I will argue, the use of reprogenetic technologies is inevitable. It will not be controlled by governments or societies or even the scientists who create it.

There is no doubt about it. For better *and* worse, a new age is upon us. And whether we like it or not, the global marketplace will reign supreme.

PART ONE

LIFE

In the beginning . . . the earth was
without form, and void, and darkness was
upon the face of the deep. . . . And God
said, "Let there be light"; and there was
light. . . . Then God said, "Let us make
man in our image, after our likeness; and
let them have dominion over the fish of
the sea, and over the birds of the air,
and over the cattle, and over all the
earth, and over every creeping thing that
creeps upon the earth." So God created
man in his own image, in the image of
God that created him; male and female
he created them.

—GENESIS 1:1–27

1

What Is Life?

I'm about to take you on an incredible journey into the future of humankind. It is a future that was unthinkable just a few years ago, beyond the reach of mortal men and women. But all that has changed, forever. We, as human beings, have tamed the fire of life. And in so doing, we have gained the power to control the destiny of our species.

This power will come from the merging of remarkable scientific and technological advances in two fields—reproductive biology and genetics—that have progressed independently of each other until now. These fields are now poised to come together as reprogenetics, and it is reprogenetics that will turn science fiction into reality, from cloning to embryo selection to genetic engineering, and beyond.

Before we explore where we can go, we need to have a sense of where we come from. We need to know exactly what life is. We need to examine how it might have originated. We need to understand how it originates over and over again with the emergence of each new human being.

Obviously, anyone reading this book is alive. It's obvious that animals, plants, and microscopic germs can also be alive. But simply using the characteristics of all earthly life forms to come up with a definition of life would be selling life short. The problem is that all living entities on earth are of a single kind—which I shall refer to as *biolife*—that is easily distinguished from nonliving things. Using biolife alone as the basis for defining life is highly chauvinistic. For if life arose independently elsewhere in the universe, the chance that it would have an appearance—either physically or chemically—similar to that of any biolife form on earth is essentially zero.

To define life in the most general way possible, we need to let our imaginations wander far away, out of the real world as we know it. A wonderful place to begin is with a science fiction story entitled *The Black Cloud*, written in 1957 by the British astronomer Fred Hoyle. The story begins with the discovery by astronomers of an extremely large, dark celestial object bearing down on our solar system at high speed. As the object enters the solar system, it begins to slow down through what appears to be spontaneous ejection of matter in the direction of its movement. The diffuse object—which has the appearance of an enormous "Black Cloud"—finally comes to a complete standstill in a position that surrounds the sun. Throughout this first part of the story, the earth's scientists succeed in determining the general structure of the Black Cloud and in explaining its movements through the well-established laws of physics.

The presence of the Black Cloud blocks most of the sun's light from reaching the earth and the death of humankind is imminent, when one scientist speculates that the Black Cloud might actually be a complex living organism that has positioned itself around our sun *on purpose* to "feed" on it as an energy source. Humans quickly make radio contact with the Black Cloud and explain their predicament. The Black Cloud is *stunned* to learn that life can exist in the form of tiny little creatures sitting on the surface of a planet, and in time, it agrees to move out of the solar system, to pursue its activities elsewhere in the galaxy.

Although the Black Cloud came to life in the imagination of a scientist, it provides a useful model for probing what it is that we mean when we say something is alive. Could a real living Black Cloud exist somewhere in our universe? Nothing that we know about science today rules it out (which doesn't mean to say that it's likely).

A different form of life, created not by a science fiction writer but by computer scientists, is referred to as artificial life, or "a-life" for short. A pioneer of the a-life field is Thomas Ray of the University of Delaware who was among the first to create a program that could reproduce itself and evolve inside a memory chip of a computer. Starting with a single copy of a program containing eighty instructions required for its reproduction and dispersion, Ray watched his computer chip world (which he named Tierra) become populated with large numbers of a-life offspring that evolved, competed, and sometimes developed symbiotic or parasitic relationships with one another.

A-life research has blossomed since 1990, and hundreds of computer scientists, sometimes working together with biologists, have created their very own a-life forms with different properties living in different types of computer chip worlds. Although these self-reproducing programs are referred to as *artificial* life, many scientists see nothing artificial about them. Rather, the a-life designation has stuck simply to distinguish them from biolife (b-life) entities.

Finally, there is HAL, one of the most infamous characters ever to appear on a movie screen. HAL was, of course, the vengeful computer that controlled the spaceship traveling to Jupiter in the 1968 film *2001: A Space Odyssey* written by Arthur C. Clarke. Although just a computer (actually a program running within a computer), HAL displayed unanticipated human emotions such as pride, anger, and fear. The proof that HAL had indeed become a virtual human being—with an internal drive toward self-preservation—came with his response to an order from one of the astronauts on the spaceship to turn himself off: "I'm sorry Dave. I'm afraid I can't do that."

THE UNIVERSAL PROPERTY OF LIVING THINGS

What properties do biolife, the Black Cloud, a-life, and HAL all share in common? What is it that *gives* life to each? Biolife and a-life both reproduce and evolve, while HAL does not. HAL, the Black Cloud, and the human form of biolife all express a reflective self-awareness while a-life and most other forms of biolife do not. Is there anything that all of these life forms have in common?

Yes. The one property that they all share is the ability to use energy to establish order out of disorder. As an 1830 quote in the *Oxford English Dictionary* says, "Life is seen in organized bodies only, and it is in living bodies only that organization is seen." All living things have a high degree of internal organization, which can also be viewed as a high level of information content. To maintain self-defining information, a living thing requires energy. When death occurs, energy use gets turned off, and the information and internal organization of the living thing begin to dissipate. Ultimately, the constant pull of entropy will cause a once-living thing to decay into the inanimate world around it.

The examples of life that we have looked at here use several different forms of energy. A-life and HAL both use the electrical energy that powers computers. The Black Cloud and plant forms of biolife feed on solar energy. Most other biolife forms are dependent, directly or indirectly, on the chemical energy that plants generate in their conversion of sunlight. However, some recently discovered biolife forms that live on ocean floors, miles below the surface, far removed from the rays of the sun, are totally dependent on chemical energy spewing forth from the earth's interior in underwater volcanic vents.

Still, just because something uses energy to create order out of disorder does not mean it's alive. In 1997, it is no longer true that organization is seen only in living bodies. Intelligent creatures like ourselves can easily create complex nonliving machines that also use energy to establish order. But if we found such anti-entropy machines on another planet, we'd have good cause to believe that there were living beings nearby with a high level of consciousness.

So if the ability to create order out of disorder does not define life uniquely, what does? In fact, we are hard pressed to find anything else that all the living systems we have discussed have in common, and yet we seem to recognize living things when they are presented to us. What is going on here?

The answer lies in the confused definition of the word life. The problem is that we commonly use this single word—life—to mean two different things. We use it to mean *life-in-general* and we also use it to mean *conscious life* in particular.

LIFE-IN-GENERAL

Before I define the attributes associated with life-in-general, I must empha-size one feature that need not be present. Life-in-general can exist not only in the absence of consciousness, but in the absence of any kind of neurological activity whatsoever. Examples of life-in-general abound on earth and include millions of different species of microbes, fungi, and plants. All are easily recognizable as alive because they are all composed of well-defined living units called cells, which we will come back to in the next chapter.

Is it possible to come up with a definition of life-in-general that is generalizable across the universe and not dependent upon the specific characteristics of biolife on earth? A definition that we could use as we search for life forms on other planets? Let's try.

First, as noted, an absolute requirement for life of any kind is the ability to use energy for the purpose of maintaining information and struc-ture. Although this property alone does not define life, it is clear that life cannot exist without it.

Second, living things generally have the ability to reproduce them-selves. *Life begets life.* There are, of course, many exceptions to the rule, such as the sterile hybrid of a horse and donkey called a mule. Some mutated forms of simple life forms, as well, can lose the ability to repro-duce. But while many individual living creatures have lost the ability to reproduce, each one is still a *product* of reproduction from a similar parent or parents. Thus, life-in-general is better defined by its past derivation— from other living things of its own kind—than by its future possibilities.

Is the combination of reproduction and energy utilization sufficient to define life in a general sense? To answer this question, let's imagine two kinds of things that humans themselves might build with just these proper-ties. Both are based on computers, but one focuses on hardware and the other on software.

The hardware version would be a sophisticated computer-driven robot that was designed with the ability to create exact copies of itself—from raw materials—without any human assistance. If solar panels were incor-porated into the robot, it would not even be dependent on a human-provided energy source. It would churn out exact copies of itself, which in turn would churn out more exact copies of the original machine, and so on, until materials needed for the manufacture of further robots were no longer available.

While a computer-driven self-replicating robot is still in the realm of fantasy, self-replicating software has already made its mark on the world in the form of computer viruses (which cause harm) and computer worms (which do not). These are small programs designed with the ability to make exact copies of themselves on the same computer disk or others, which may be connected through networks over distances of thousands of miles.

Is a self-replicating machine or a self-replicating program alive? Most people would say no, even though they may attach cute appellations like worm or virus to the programs. There is something missing, some essential ingredient not present that we do find in living things. The essential ingredient is the single feature that distinguishes the artificial-life programs discussed earlier from simple self-replicating programs like computer viruses. Scientists think a-life programs are alive because they have the ability to *evolve* new properties. Any computer scientist could write programs that *simulate* features of living biological communities such as symbiosis and parasitism. But to watch these new properties emerge *spontaneously* in the offspring of programs that did not express them is a different thing altogether.

Evolution can occur only when the copying process is *not* exact, when the offspring of programs or machines sometimes differ slightly from their progenitors, and when the difference is one that can be passed on to future generations. Imagine what might happen if our self-replicating robots could copy themselves with occasional self-perpetuating mistakes on a world where human beings ceased to exist. Eventually, the raw material needed to make new robots would become scarce, so that there was only enough for some, but not all, to reproduce. Because of the inexact copying process, individual robots would no longer operate in exactly the same way. One might be able to dig deeper for hidden material, another might have gained the ability to cannibalize its cousins for needed parts. These particular robots would be able to reproduce, while those that had remained unchanged would not. It wouldn't take many generations before robots with enhanced properties were the only ones left, and these would continue to evolve in response to both inter-robotic competition and ever-changing environmental conditions. Are there any limits to the future forms that these evolving robots might achieve? I don't think so.

It seems clear that once the process of reproduction is coupled with the ability to evolve (through survival of the fittest), life-in-general exists. In fact, evolution alone may be considered the overarching theme around which life must be organized since, by necessity, it incorporates both reproduction and the use of energy to maintain and generate information and organization. True, evolution does not take place within any individual organism. But it does seem possible to provide a complete definition of life-in-general at the individual level as a *product* of reproduction and evolution that uses energy to maintain self-defining information and organization. The inanimate becomes animate only upon achieving the ability to evolve.

The conceptualization of life that I have described here emerged from the grand synthesis of evolution and genetics that took place during the middle of the twentieth century. Among the theoretical biologists who played a major role was Theodosius Dobzhansky who coined the aphorism, "Nothing in biology makes sense except in the light of evolution."

One final feature of life, *at least as we know it*, is complexity. There seems to be a minimum level of complexity, a minimum number of interacting components working toward a common goal, that is required for a living thing to survive, reproduce, and evolve against the constant pull of entropy toward decay and disorder. Even the smallest living cell contains millions of complex molecules undergoing billions of predefined interactions with one another. The minimum level of organized complexity at the core of living things poses serious problems for explaining how it all started, as we will discuss in the next chapter.

LIFE IN A SPECIAL SENSE

Now it's time to consider the very separate meaning that we give to *conscious life*, in its human form. Let's begin with a thought experiment. Let's imagine that not only has artificial intelligence become a reality, but that it can be placed into synthetic bodies that are indistinguishable from human beings. Now let's assume that you discovered one day that your best friend was not really made of flesh and blood as you had thought, but had been created from electronic components instead—like an intelligent version of one of the Stepford Wives or a compassionate version of the Terminator. How would this discovery change the way you felt about your best friend? Would you suddenly terminate your relationship? Would you no longer care if this person was dead or . . . alive?

What is it about your electronic friend, or our old friend HAL, that makes them alive? They are clearly not products of reproduction and evolution (from their own kind), nor do they have the ability to reproduce themselves. Instead, what makes us view them as alive is their ability to feel and express a range of genuine human emotions and, most important, their attainment of the uniquely human condition of reflective self-awareness.

So we now see that there are two very different meanings of the word *life* as it is used in connection with humanness. One meaning is associated with the basic processes of energy utilization, maintenance of structure and information, reproduction, and evolution that are shared by all living things. In the context of biolife specifically, *life in a general sense* is rooted within the individual cell, which will be the focus of our attention shortly. In contrast, the second meaning of the word *life* is rooted within the cerebral functioning that gives rise to consciousness. In human beings, *life in a special sense* is localized to the region between our ears, but it lies far beyond the level of any individual nerve cell.

Although for the sake of simplicity, I have limited the discussion here to the two meanings of life that arise in the context of human beings, the same dichotomy exists for all animals with a nervous system, from microscopic soil worms to chimpanzees. It is the nervous system of each animal, large and small, that is responsible for the characteristic behaviors that define that animal as a whole. An animal dies— *in the special sense* of the word—when its nervous system ceases to function, even though most of the individual cells within its body remain alive *in the general sense* for a longer time.

The difference between *life in the special sense* and *life in the general sense* is well illustrated by two things that can happen after a person dies (when the heart stops beating and brain activity has ceased). If the person was an adult man, his hair follicles will continue to function, giving rise to an unshaven face with an extra day's growth of beard on a dead body. A dead body can also be mined for living organs that can be used for transplantation to save the lives, *in the special sense*, of other people. People who have received heart, lung, or kidney transplants do not have a personal identity that is any different from the one they had before their operation, even though they are now a mixture of two different living systems, *in the general sense*.

A problem with the definition of "life in the special sense" that appears to be unavoidable lies in the difficulty of describing what we mean by a *functional* nervous system. At what stage during fetal development does a nervous system become functional? And at what point is a neurologically incapacitated adult no longer alive?

Any answers to these questions will be highly subjective. And I doubt that further scientific understanding will ever make the task easier. As a result, there will always be a large fuzzy zone at the borders of *life in the special sense*. But the absence of a sharp line does not invalidate the definition; it just instructs us of its limitations.

The inability to separate the two different meanings of the word *life* can cause confusion when people talk about whether or not a fertilized human egg is alive. According to our general definition of life, the fertilized egg is obviously alive in the same sense that cells are alive in donated blood or donated organs. But, this single cell does not represent *human life in the special sense*. Whether the human embryo should be treated in a special way because it contains human genetic material or because it has the *potential* to develop into a *human life*, is a question we will return to in chapter 3.

2

Where Does Life Come From?

THE CELL IS ALIVE

The basic living unit of biolife is the cell. All creatures large and small—from the blue whale and the giant sequoia to each of the billions of symbiotic bacteria that live inside your gut helping you digest your last meal—are composed of one or more of these microscopic entities. If you break a cell apart, the individual pieces that you obtain are no longer alive. They are just built-up molecular structures—combinations of atoms—and nothing more. Some cell fragments may be able to utilize energy to maintain their structure for brief times under special conditions, but they can't reproduce themselves. Thus, biolife cannot be reduced to any unit smaller than a cell, and the simplest cell possible is still inordinately complex.

It was once thought that the laws of physics and chemistry alone were not sufficient to explain the notion of aliveness that is associated with a cell. It was assumed instead that a special *vital force* acted upon the molecules within a cell to bring them to life. We now know that there is no vital force and that all of the molecules within the cell obey the inanimate laws of physics and chemistry. But the cell *is* more than just the sum of a large number of molecules. Together, the complex interlocking networks of molecular interactions that occur within the cell yield the emergent property that we call life. If you try to take it apart, you just get molecules; put it back together, if you could, and you have life.

Animals and plants grow larger not by increasing the sizes of their cells, but by increasing their cell number. The liver cells of a human, for example, are indistinguishable from the liver cells of a mouse. But the whole liver of an adult human has a thousand times as many cells as the whole liver of an adult mouse. All together, each adult human being has about 100 trillion cells in her or his body, which is about 100 trillion more than the simplest free-living creatures composed solely of single cells.

The microscopic size of the cell should not lull you into the belief that it is a simple object. While we may not be able to see a cell with our eyes alone, it is still gargantuan relative to the size of its individual molecular components. The workings of each individual cell within the human body are, at a certain level, just as complex as the communications that take place among our cells that help to define us as human beings.

With microscopes and the tools of biochemistry, a cell can be recognized by both its appearance and its inner workings. It is surrounded by an ultrathin skin called a plasma membrane. Within the confines of its plasma membrane, the cell is like an exquisite piece of machinery with hundreds of thousands of working parts, each localized to a specific compartment and each communicating with multitudes of other cellular components. So long as we are alive, the machinery in each of our cells never shuts down or goes to sleep. Even when *we* are asleep, each of our cells is constantly churning—a veritable beehive of activity—burning up calories.

The information required to produce every one of the cell's many components in the right numbers and to place them all in the right places is encoded within its genetic material, DNA. Furthermore, all of the information required to build complex organisms—like human beings—with large brains capable of conscious thought is also encoded within molecules of DNA.

Amazingly, the DNA present in a conscious human being is the same as the DNA present in a tiny amoeba. The reason these two organisms look different from each other is because the *messages* that they carry in their respective DNA molecules are different (and because each of our cells has a thousand times as much DNA as an amoeba). Indeed, the unique characteristics of every species on earth are all a consequence of unique messages recorded in a common molecule.

All cells have two separate compartments called the "cytoplasm" and the "nucleus." The nucleus has its own membrane and sits like a ball in the middle of the cell. It contains all of the genetic material within structures called chromosomes. Single-cell organisms can carry just a single chromosome, while normal human cells carry 46. Each chromosome contains a single DNA molecule.

All the cellular material that lies outside the nucleus, and inside the plasma membrane, is called *cytoplasm*. The cytoplasm contains the machinery that interprets genetic information flowing from the nucleus and responds to it by building all the structures that make up the cell. The cytoplasm also relays signals from the outside world—meaning all the other cells in the body as well as the external environment—into the nucleus where specific changes in the program of gene expression can be elicited.

There are two steps in the process of cell reproduction. Cells must generate more of their component parts as they increase their size by twofold. They must also make accurate copies of each of their DNA molecules. When both of these processes are completed, cell division can occur. At the completion of cell division, there are two "daughter cells," each containing its own copy of the complete genetic material that was present in the original "parent cell" (which ceases to exist). For single-cell organisms, cell division is equivalent to reproduction. In contrast, multicellular organisms use cell division to grow in size and complexity.

THE ORIGIN OF LIFE ON EARTH

To a biologist, the similarity of all living things on earth to one another is much more striking than any differences that might exist. Not only are all living things made up of cells, but the cells of all living things work in essentially the same way with the same complex molecules and the same type of genetic material, which is read according to the same genetic code. There is no fundamental law of biochemistry that says all living cells have to be constructed in the particular way that they are. A smart biochemist could imagine an almost infinite number of other ways in which to build a functioning cell that, like biolife, is based on chemistry.

Furthermore, even if one did start out with a cell like the one that defines biolife, there is absolutely no reason why the genetic code should be as it is. Imagine that you have grown up speaking the English language without ever learning how to read or write. Suddenly, someone gives you a set of twenty-six symbols and tells you to use these to invent a written code for what you are speaking. Let us say that the twenty-six symbols happen to be the twenty-six letters of our alphabet. What are the chances that you will use the symbol *M* to represent the starting sound of the word you use to address the woman who raised you? The answer is one chance in twenty-six. Now what is the probability that you will correctly assign every symbol in the alphabet to its appropriate sound (assuming for simplicity that you start out with the twenty-six sounds that English-speaking people happen to assign letters to). The probability is 1 chance in 403,291,146,110,000,000,000,000,000, or as a scientist would write it, approximately 1 chance in 4×10^{26}. As far as we're concerned, this is essentially zero. Although the probability is somewhat different for the genetic code (it's actually smaller), the same principle holds. There is an infinitesimal probability that two cells would use the same genetic code *just by chance* even if they did both happen to use DNA as their genetic material. And yet, every living animal, plant, and germ cell on earth uses the same genetic code.

The inescapable conclusion is that all living things on earth—all animals, plants, and microorganisms—are descendants of the same original cell that happened to begin life with the particular genetic code that scientists now refer to as *universal* (although *global* would be more accurate). The first cell existed for a fleeting moment some 3.5 billion years ago, and it represented the beginning of life as we know it. Quite soon after its virgin birth, the original cell divided into two, both of which soon divided again and again, and within an instant of geological time, the earth was covered with single-cell organisms. If an alien were watching the earth from afar and happened to blink, she would have missed it—a barren planet transformed to one suddenly teeming with life. Only simple life forms for sure, but powerful enough together to change the physical nature of the planet and the chemical nature of its atmosphere.

There may seem to be an enormous gulf between a single-cell organism—so small that we can't even see it—and a human being composed of 100 trillion cells, but no true biologist has any doubt that one evolved into the other. Our confidence is based not just on what Darwin's theory says *should* happen, but on our ability to see the critical intermediate stages of evolution, all the way up the ladder of complexity from single cells to sponges, to worms, to fish, to reptiles, to mammals, to primates, and finally, to us.

We see these intermediate stages both in the fossil record as well as in the living fossils that are all around us. It's the fossil record that allows us to date the initial appearance of each intermediate stage. But it's the living representatives of each stage, and our ability to study them with the tools of molecular biology, that give us a glimpse into how evolution took place, step-by-step. Scientists are lucky that so many critical intermediate forms of biolife kept on reproducing in a primitive state, even as their cousins went on to evolve to the next stage of complexity.

In contrast, there are no naturally occurring intermediates between very simple molecules and the organized complexity of the cell. Thus, any attempt to define the intermediate steps that led to the construction of the first cell must be purely speculative. Since most biologists believe that life had to coalesce spontaneously from inanimate matter on earth, there is no limit to such speculation.

There are two assumptions that go into any speculation about the origin of life on earth. The first is that highly improbable events become not only probable but likely within a long enough time, and the 1-billion-year period between the hardening of the earth's crust and the appearance of the first cell is certainly long. The second is that intermediate stages leading up to the first cell did exist, but they've all disappeared without a trace.

The first assumption can be illustrated by analogy to the chances of buying a winning ticket to a daily million dollar lottery. Let's say that every day, 2 million people each buy one ticket, and every day, one person is chosen as a winner. What is the probability that you will win the lottery on any particular day? The answer is 1 chance in 2 million, which is about the same as the probability that you will be hit by lightning during the course of any particular year. But, now let's say that you buy 1 lottery ticket every day for 9 million days (remember, this is a thought experiment). What is the probability that you will win at least once? Well, it doesn't matter if you bet the same number each day or a different number, the answer is a remarkable 99 percent. This thought experiment shows how a highly improbable event can turn into an extremely likely event given enough throws of the dice.

But, statistical tricks allow us to go only so far. As the initial probability becomes smaller and smaller, the time required to make it likely becomes longer and longer. And the probability that individual molecules will coalesce spontaneously into even the simplest cell imaginable—with interlocking networks of metabolic and genetic activity—is so low that we are justified in saying it could never happen during the entire time span of the universe.

It is for this reason that most biologists assume life got a toehold on earth through a series of simpler intermediate stages of molecular existence and that each stage provided a platform for evolution into the next. Each stage would have to represent a proto–life form able to reproduce itself into a large enough number of copies so that a highly improbable event could occur in one and move it up to the next stage and so on, until ultimately, the first cell appeared.

The first step on the pathway to life must be a molecule that can make copies of itself. In *The Selfish Gene*, Richard Dawkins christens such a molecule the "replicator." Although we can only speculate about the nature of the original replicator, it must have been simple enough to form spontaneously (given a billion years of trying) from atoms crashing into each other on the surface of the primordial earth. And once it was formed, its primary distinguishing characteristic would be its ability—all by itself— to cause the formation of other molecules *in its own image*.

What was the first replicator? Until recently, biochemists believed that it was an earlier variation on a molecule that all modern-day cells use. This point of view was supported during the 1950s with the results of a fascinating experiment that showed how the fundamental building blocks of proteins—amino acids—were created spontaneously under laboratory conditions thought to simulate the environment of the early earth. This finding led to early speculation that the first replicators were protein-like in nature. But no matter how hard they tried, scientists couldn't imagine how protein-like molecules could make copies of themselves.

In the 1980s, it was discovered that a molecule called RNA, a close relative of DNA, not only has the ability to carry genetic information (like DNA), but also has the ability to perform chemical reactions. This discovery caused a massive migration of scientists to the belief that RNA was the original replicator, and that the cellular apparatus built itself up around it. However, many thoughtful scientists are not happy with the RNA molecule as the original replicator for the simple reason that it is not simple enough. The structure of RNA is so complex that even if it did appear on the earth sporadically over the first billion years, it wouldn't have had the capacity to make copies of itself, all by itself.

And so in recent years, chemists have suggested that biologists may have been too shortsighted in their attempts to discover the original replicator as a primitive form of something that is still present within modern-day cells. Perhaps, they said, the original replicator was something entirely different. Perhaps, DNA, RNA, and proteins didn't appear until much later in the game of life. But once they *did* appear, evolving out of the original replicator or its offspring, they would easily out-compete their progenitors, and the origins of life would disappear from the earth.

In a delightful book called *Seven Clues to the Origin of Life*, A. G. Cairns suggests that the original replicator might have been an inorganic crystal, of a type found in clay or mud. Crystals are defined by their ability to grow spontaneously, reproducing whatever structure they start with. Under the right conditions, a crystal could expand in size, then break apart into two or more smaller crystals that could then expand in size again.

We don't normally see crystal growth as evidence of life and, indeed, in most cases we shouldn't. But what our lottery example tells us is that improbable events can become probable over a long enough time. Thus, it is possible to propose that particular crystals could have incorporated more sophisticated molecules from the environment that provided a better chance of survival and propagation, and evolution could have taken off from there. The distance from simple crystals to a complete cell is still enormous and unexplained, but the crystal birth of life is surprisingly compelling.

Although most biologists insist that life had to have originated spontaneously on our planet, other ideas have been considered. One alternative solution to the problem of missing intermediates was proposed early in the twentieth century by the Swedish chemist Svante August Arrhenius. In 1908, he suggested in his Panspermia (meaning *seeds everywhere*) theory that life originated elsewhere in the universe and was carried to earth by spores floating freely in outer space. The Panspermia scenario is analogous to the mechanism by which a newly formed volcanic island, thousands of miles away from other land masses, will become seeded with new animal and plant life blowing onto its shores with the waves and winds of enormous storms—an example of the improbable becoming very likely over a sufficient time. Although it's an interesting idea, scientists find no evidence of freely floating life forms in outer space.

Francis Crick, one of the co-discoverers of the structure of DNA, has tried to rescue Panspermia by elaborating it into a theory of "Directed Panspermia," which presupposes that spores did indeed land on earth, but that they were sent here deliberately in a spaceship by a highly advanced, but doomed, civilization living elsewhere in our galaxy. The major redeeming feature of this or any other form of the Panspermia theory is that it solves the problem of missing intermediates on earth. These theories also provide an opportunity for the spontaneous generation of cellular life on a planet with special environmental conditions that would have been more auspicious than those found on the early earth. But what these *better-than-earthlike* conditions could possibly be is not at all clear.

There is another solution most biologists tend not to consider: Divine Intervention. With Divine Intervention, the first cell was indeed put together deliberately—in the same way that a team of Boeing workers might build a 747—by a Supreme Being that people call God.

Divine Intervention is generally rejected as a solution to biological problems because not only is it not required, but there does not seem to be any room for it after the first cell appeared. The 3.5-billion-year-long pathway of evolution to humankind can be explained by Darwinian principles of natural selection and nothing else. Furthermore, there is no evidence for any violation of the basic laws of physics and chemistry in any biological process that operates inside a living cell. And to many scientists, it would seem that Divine Intervention requires these laws to be violated by an invisible hand that intervenes, at least in certain situations at certain times.

But, even if one completely accepts this point of view, it is still possible to argue that the birth of the first cell is the one event that shows the hand of God at work, and that after the one act of creation, the Supreme Being interfered no more. There's an added assumption that goes hand-in-hand with this particular version of Divine Intervention. That is, once the first cell was in place, the evolution of human beings was inevitable. It is the notion of inevitability that was behind Pope John Paul's decision to accept evolution as an established scientific theory.

Interestingly, if we were able to time-travel back to the moment at which the first cell appeared, we would not be able to distinguish the single-event version of Divine Intervention from Directed Panspermia. In both scenarios, something or someone outside the earthly realm consciously created the first cell on our planet, and then left it alone. With both scenarios, however, a critical question still remains: How did the earlier Being, or one even earlier than it, come into being?

There is another serious problem with any benign neglect theory of human inevitability that lies squarely within the process of evolution itself, a problem that has been addressed most forcefully by Stephen Jay Gould. The word that Gould uses so often to describe the history of life is *contingency*. No matter what environmental conditions existed on earth throughout the millennia, the life forms that happened to appear could not have been anticipated. This is because there is no single solution to any evolutionary problem. At each moment in the past, one out of many solutions was chosen *just by chance* in each organism that survived. Billions of chance events, one after another, none that had to occur, finally led to us. Gould is fond of a thought experiment in which we "re-wind the tape" and allow life to evolve again, from any early point, in the same precise environment over the course of earth's entire history. The probability that creatures with humanlike consciousness would appear a second time is virtually nil.

The conclusion reached by Gould and other scientists is disturbing: We are an accident. Given the smallest change in the wind, some other creature would have won the lottery. For every genetic change made in response to predators or changing environmental conditions, thousands of others could have succeeded just as well. If the asteroid that hit the earth and killed the dinosaurs 65 million years ago had shot past our planet instead, mammals would never have come to replace them as the dominant life form, and without our ancestors, we would not be here. As the Nobel Prize–winning physicist Steven Weinberg says at the end of his book describing *The First Three Minutes* of the universe, "The more the universe seems comprehensible, the more it also seems pointless."

Not all scientists feel this way. Freeman Dyson, one of the founders of modern quantum theory, suggests that the uncertainty inherent in the quantum mechanical view of the world could provide a hidden means for Divine Intervention into the process of evolution that does *not* break any laws of physics or chemistry. "Matter is weird stuff," Dyson says, "weird enough so that it does not limit God's freedom to make it do what he pleases."

Other physicists have pointed out a remarkable coincidence in the fundamental physical properties of the universe. It turns out that if the masses of any of the fundamental particles—like quarks and electrons— or the forces between them, or perturbations in the early universe, were different in even the smallest way, the universe would not have been able to produce galaxies; galaxies would not have been able to produce stars; and stars would not have had orbiting planets covered with carbon atoms and other elements required to produce the molecules of life.

This remarkable finding has been used by some physicists as a back-hand way to explain the particular properties of our universe. The logic— known as the Anthropic Principle—goes like this. If the universe had any *other* properties, there would be no intelligent life around to see it. Since there is intelligent life, the universe has to have the properties that it has. In other words, the properties of our universe are explained simply by us being here to ponder them.

Although the word *universe* is commonly used to mean "the totality of all existing things," some physicists use the word to mean just the totality of all space that is *physically connected* to the space in which we live. What Einstein's General Theory of Relativity suggested is that multiple, finite universes could coexist unbeknownst, and unconnected, to one another. This means that there might be many, many universes that are each characterized by different fundamental physical properties. By chance, our universe has properties that support the emergence of life and that is why we *could* be here, although we did not *have* to be here. According to this point of view, there need not be a God. In the near-infinitude of existence, life happened by chance.

Ed Witten, a leading practitioner of modern-day theoretical physics, rejects this version of the Anthropic Principle. Witten believes that the fundamental physical properties of our universe—and all universes—may someday be explained *directly* by a new theory of physics that brings together the currently separate theories of general relativity and quantum mechanics. If a final theory of physics explains it all, then it will no longer be possible to ascribe the special life-sustaining properties of the universe to chance alone. Instead, it will seem as though the universe was con-structed *on purpose* in this manner, although the meaning of "on purpose" will still be subject to debate.

One philosophical interpretation is a version of "theism" with a benev-olent God who left the picture long before the first cell was born. After creating the universe and simply setting it into motion at the moment of the Big Bang, with special physical laws that guarantee the emergence of conscious beings somewhere, this God sat back to watch life unfold as isolated sparkles throughout his domain. Freeman Dyson goes even further by suggesting that "the laws of Nature and the initial conditions are such as to make the universe as interesting as possible [to intelligent creatures like us]."

An alternative philosophical interpretation of an all-inclusive final theory of physics is an "emergent" version of "pantheism." Pantheism is a philosophical doctrine that considers the universe and God as a single, indivisible entity. In its emergent version, each conscious mind would represent a "small piece of God's mental apparatus," with the whole evolving and maturing over time. In this view, God may be seen as the "Created" rather than the "Creator."

Unfortunately, each explanation proposed for the origin and meaning of life and the universe is subject to serious philosophical predicaments. Theories that posit the existence of a Creator of any form fail to explain where this Creator came from and who its Creator was. Theories that posit the accidental creation of consciousness fail to explain how the universe came into being and *why there is something rather than nothing.*

Are there answers to be had to the great questions of our existence? And if so, will humankind ever learn what they are? My own feeling is that there are answers, and we will eventually have them. We have come to understand much about life and the universe in a remarkably short period of our history. And we are now poised to use reprogenetics together with other technologies to expand our powers to probe the universe in ways that we can only begin to imagine today. It is for this reason that I refuse to believe that knowledge exists that is beyond our reach.

For the moment, though, let us summarize what we do know about the history of life on earth alone. We know that the earth was a *barren* place when its surface cooled down and solidified 4.6 billion years ago, and that a billion years or so later, there were single-cell organisms alive and reproducing. It seems probable that cellular life began here just once. If other independently derived forms of life existed at any time, they have disappeared without a trace. And so it is that the first cell divided into two and each of these divided into two as well. And some of these daughter cells died off while others changed in response to their environment and continued to divide. And after a hundred billion or so divisions, in an unbroken line, *the* first cell gave rise to *your* first cell.

3

Does Your First Cell
Deserve Respect?

We all have at least a general understanding of how babies come into existence. But although a general picture of reproduction is clear enough, we often miss or misunderstand subtle but important facts. Thus, it is worthwhile to take a closer look at "the facts of life" as they pertain to the critical events that occur immediately before and after the formation of a fertilized human egg. With the relevant scientific information in hand, we will be in a position to consider an important ethical question with implications for most of the reprogenetic technologies that will be discussed in this book: What is the moral status of the human embryo?

SPERM AND EGGS

Fertilization occurs when a sperm cell and an egg cell join together, becoming one new cell in place of the two cells that existed previously. To appreciate the process of fertilization, we must first understand the special properties that distinguish sperm and egg cells (both also known as *germ cells* or *gametes*) from every other cell in the human body.

All your other cells (referred to as *somatic cells* to distinguish them from germ cells) carry the same genetic material distributed across twenty-three pairs of DNA molecules stored within twenty-three pairs of chromosomes. One member of each chromosome pair came from your mother and the other member came from your father. The two members of any chromosome pair are 99.9 percent equivalent to each other (with one exception). This represents an enormous amount of redundancy, as if each cell carried the written text from two editions of the same large encyclopedia, with each edition differing only in one word on every other page.

To extend the metaphor further, it is the overall similarity between all possible editions of the encyclopedia—known generically as the human genome—that is responsible for the multitude of ways in which all human beings are the same as all other human beings. On the other hand, it is the small differences between editions that make you look and, to a certain extent, behave differently from other people. Who you are, in particular, depends on the ability of all your cells to combine the information present in the two editions of the human genome encyclopedia that you received from your father and mother, respectively.

Each of your sperm or egg cells contains only a *single* edition of the human genome within just twenty-three chromosomes. But in no case is the single edition the same as the one you received from either your mother or your father. Instead, early in the process that leads to the production of each individual gamete, your maternal and paternal editions of the human genome exchange random pages and whole chapters with each other *in a very precise way* so that entirely new editions of the encyclopedia emerge. The new editions have all the same chapters as before, but each is a random mixture of your mother's and father's genetic material. And only one of these editions makes it into each of the individual sperm or egg cells formed in your body at the end of the process.

Every sperm cell, of the billions produced during a man's life, and every egg cell produced by a woman bears a different composition of genetic material, a different mixture of human genome editions. It is for this reason that, with the exception of identical twins, it is impossible for a human couple to have two genetically identical children.

CONCEPTION

Millions of immature eggs—called *oocytes*—are stored in a woman's ovaries. Every month or so, during the fertile period of a woman's life, hormonal signals cause one of these oocytes to ripen into an egg that is capable of being fertilized. When the ripening process is complete, the egg is released from the ovary—an event referred to as ovulation. Upon leaving the ovary, the egg begins its slow journey down the fallopian tube where it remains receptive to sperm for about 20 hours. If this brief opportunity is missed, the egg will degenerate and pass from the fallopian tube to the uterus and out of the body along with the uterine lining at the time of menstruation.

If sexual intercourse occurs within a day before or after ovulation, sperm will make their way up toward the egg in the fallopian tube. Human sperm probably have no ability to actually seek out the egg, which is why men ejaculate 100 million or more of these gametes, when all but one—at most—are destined to die. Through the sheer force of numbers, some sperm will, by chance, make contact with the egg.

The egg cell itself is surrounded by a rubbery coat called the *zona pellucida,* or zona for short. The egg is like a ball loosely floating in fluid within the hollow sphere of the zona. When a sperm cell bumps into the zona surrounding an unfertilized egg, the two stick together. The zona now induces the front end of the sperm to release a concentrated essence of digestive enzymes, and with the forward motion caused by its wagging tail, the sperm follows a tiny burn path through the zona into the fluid between the zona and the egg cell. The first step of fertilization is now complete.

A handful of sperm cells will often make their way into this isolated space, still swimming about. By chance, one will be the first to stick to the actual egg cell membrane. When this happens, the sperm and egg cells begin the process of fusion. During the process, a small portion of the egg's membrane actually surrounds the tiny sperm cell and gulps it in. The sperm cell remains intact and swimming, at first, inside the egg's cytoplasm. But, within a few minutes, its tail and the membrane surrounding its head begin to disintegrate. The second step of fertilization is now complete.

The fusion of egg and sperm triggers rapid responses to prevent the entry of other sperm into the egg, including a hardening of the zona coat, so that additional sperm can't bore through, and an electrical screen around the egg membrane so that sperm already inside the zona are repelled by the egg's surface. These events are, of course, not instantaneous, and occasionally, a second sperm will slip in during the seconds that it takes to raise the barriers. Such double-fertilized eggs are overloaded with genetic material and cannot develop properly; they will die within several days.

Fusion also sets the fertilized egg—now called a *zygote*—on the slow but steady course of embryonic development. And one of the first tasks that the egg undertakes is a reduction of the mother's genetic material by half. Are you confused by this last statement? Didn't the reduction of genetic material occur back in the ovary before ovulation? Well, actually no. The process *began* in the ovary; in fact, it began before the mother was even born (and we'll return to the bizarre implications of this fact in chapter 14). But it was never completed—an irksome scientific fact with implications that we will consider in a few moments.

Meanwhile, the nucleus that was contained within the tiny compressed sperm head is undergoing a slow expansion to become the same size as the nucleus contributed by the egg. These two nuclei are actually referred to as *pronuclei* by scientists because each contains only half the genetic material found in the normal nuclei of somatic cells. Thus, they are not full nuclei, just the precursors to a full nucleus.

Contrary to popular belief, however, the two pronuclei never fuse into one. Instead, throughout the one-day life of the zygote, the genetic material provided by mom and dad remain cloistered in their own separate spheres. How can this be true? In almost every popular book and article that discusses fertilization, there is some statement about the fertilized egg having a *single* nucleus containing the chromosomes of both the mother and the father together for the first time. In fact, such a thing has never been seen. It seems likely that this false concept began as wishful thinking (for reasons we will return to in a moment) that was propagated from one writer to another.

What actually happens is that the chromosomes in the two pronuclei duplicate themselves separately, and then copies from each come together inside the actual nuclei formed *after* the first cell division. It is within each of the two nuclei present in the two-cell embryo that a complete set of forty-six human chromosomes commingle for the first time. Fertilization is now complete.

Now that it has two cells, what do we call it? The terms *zygote* and *fertilized egg* are used synonymously to describe only the single cell stage, which ends with cell division. Actually, animal biologists use the term *embryo* to describe the single cell stage, the two-cell stage, and all subsequent stages up until a time when recognizable humanlike limbs and facial features begin to appear between six to eight weeks after fertilization. From that time until birth, the term *fetus* is used. The word *conceptus* is also used as an inclusive term to describe all stages from fertilization to birth.

You may have noticed that I said *animal* biologists use the term embryo. I've made this distinction because a number of specialists working in the field of *human* reproduction have suggested that we stop using the word *embryo* to describe the developing entity that exists for the first two weeks after fertilization. In its place, they proposed the term *pre-embryo*. From now on, the word *embryo* was to be reserved for the group of cells that emerge at about fourteen days after conception and go on to produce the fetus rather than the placenta, as I will describe in a following section.

It is rare that scientists change well-established terminology, and such changes normally occur only in response to new scientific understanding that invalidates the use of earlier terms. Yet our scientific understanding of early embryonic development has remained essentially unchanged for more than half a century. So why is there suddenly a need to adopt a new word?

I'll let you in on a secret. The term pre-embryo has been embraced wholeheartedly by IVF practitioners for reasons that are political, not scientific. The new term is used to provide the illusion that there is something profoundly different between what we nonmedical biologists still call a six-day-old embryo and what we and everyone else call a sixteen-day-old embryo.

The term pre-embryo is useful in the political arena—where decisions are made about whether to allow early embryo (now called pre-embryo) experimentation—as well as in the confines of a doctor's office, where it can be used to allay moral concerns that might be expressed by IVF patients. "Don't worry," a doctor might say, "it's only pre-embryos that we're manipulating or freezing. They won't turn into *real* human embryos until after we've put them back into your body."

Biologically speaking, an important developmental event does occur at fourteen days. But there are other important developmental events that occur before that time and many more that occur later. The relative significance of these events has been aptly described in the final report presented to the British Parliament by the first committee ever commissioned by a government to look into new reproductive technologies: "There is no particular part of the developmental process that is more important than another; all are part of a continuous process, and unless each stage takes place normally, at the correct time, and in the correct sequences, further development will cease."

DOES THE FERTILIZED HUMAN EGG REPRESENT HUMAN LIFE?

Although life has been continuous since the origin of the first cell, you are clearly a distinctly individual human being. When did this distinction occur? And when did you—as a person—first become alive in *the special sense* that we give to the word? These simple little questions and the various answers that people give to them are at the heart of a political storm that has polarized American society like no other during the last quarter of the twentieth century. The storm I am speaking of, of course, is the issue of abortion access, and the battle lines are drawn between those who call themselves "pro-choice" with an emphasis on the pregnant woman, and those who call themselves "pro-life" with an emphasis on the fertilized egg, embryo, and fetus.

Much of the debate centers on what is referred to by bioethicists as "the moral status of the embryo." If we can figure out how the embryo should be viewed in relation to a human being, we might be able to decide how it should be treated. If we decide that the embryo is deserving of the same protection as a child, we might want to think carefully not only about abortion, but about the kinds of embryonic manipulations in the laboratory that will be discussed in this book. If we decide that an embryo is not a child, then what is it? Is it a protochild, still deserving of some special thoughts, but not quite like a child? Or is it simply a clump of human cells, no different from those we wash off our skin with soap and water?

In this section, I will be using the term embryo to refer to only the earliest stages of development that occur within a few days after fertilization. During this period, the embryo really does look like nothing more than a clump of cells. An ethical discussion of later stages of development will be left to the sections that follow.

While many philosophers and scientists have written about the status of the embryo, their opinions have not strayed far from one of the following three points of view: At one extreme are those who say the embryo is equivalent to a human being. This belief implies that embryos should be given the rights, protections, and respect that we give to all other human beings. This is the current position of the Catholic Church and many others who place themselves at the pro-life end of the political spectrum.

At the other extreme are those who say that the embryo is no different from any other clump of human cells and should not be treated in any special way. Most contemporary biologists would probably place themselves in this camp.

The third point of view hovers between these two extremes and has been stated succinctly by the noted reproductive ethicist and lawyer John Robertson: "The embryo deserves *respect* greater than that accorded to other human tissue, because of its potential to become a person and the symbolic meaning it carries for many people. Yet it should not be treated as a person, because it has not yet developed the features of personhood . . . and may never realize its biologic potential." As Robertson notes, this is probably the most widely held view among secular bioethicists.

Before joining the debate, we need to establish some basic facts through a series of questions and answers:

1. Is the embryo *alive*? Clearly, yes.
2. Is the embryo *human*? Yes again, but so are the cells that fall off your skin every day.
3. Is the embryo *human life*? No. Recall from the first chapter the two different meanings that we give to life—one for life *in the general sense*, and the other for human life *in the special sense*. The embryo does not have any neurological attributes that we ascribe to human life in the special sense.

Given that early human embryos are not alive by the definition developed here for human life, how could anyone equate them with a human being? The answer I hear most often goes something like this: First, the genetic constitution of the embryo is new and unique. Second, the embryo has the potential to develop into a full-blown human being. Taken together, the implication is that the embryo has the potential to become not just any human being, but a unique human being that is defined already by its unique composition of genetic material, and this potential does not exist within the unfertilized egg or sperm. With the powerful methods of genetic analysis that will be discussed in chapter 17, it might even be possible to get a sneak preview of what the person will look like and what his or her temperament might be. And nothing should stand in the way of the *natural* development of the embryo into that already molded human being.

The Vatican says: "modern genetic science brings valuable confirmation. It has demonstrated that, from the first instant, the program is fixed as to what this living being will be . . . recent findings of human biological science . . . recognize that in the zygote (the cell produced when the nuclei of the two gametes have fused) [original parenthetical phrase] . . . the biological identity of a new human individual is already constituted."

And the noted University of Chicago bioethicist Leon Kass argues: "While the egg and sperm are alive as cells, something new and alive *in a different sense* [original emphasis] comes into being with fertilization . . . there exists a new individual with its unique genetic identity, fully potent for the self-initiated development into a mature human being . . . Any honest biologist must be impressed by these facts."

Should we be impressed? It seems like a powerful argument. But, before we get carried away, let's ask two critical questions: Is what they are saying true? And what are they *really* trying to say?

Let's start by analyzing the scientific validity of the argument. First, as we now know, "the biological identity of a new human individual" is not constituted at the time of zygote formation. A few hours must still pass before the zygote is able to rid itself of half of the mother's DNA.

Second, the assertion by the Vatican and most others that the zygote is "the cell produced when the nuclei of the two gametes have fused" is also wrong. What might seem like a minor scientific detail to biologists is actually a critical component of the argument used by some to claim that the zygote is alive "in a different sense." People have been led to believe that the genetic material of the mother and father "come together" in a form of molecular marriage, when in truth, they remain celibate. If commingling of parental DNAs were really the deciding factor, then life would begin at the two-cell stage and not in the fertilized egg. But even at this stage, DNA molecules from the mother and father do not actually interact or even *touch* each other, they're just closer together than they were before. As a molecular geneticist, however, I cannot see why the precise distance that separates DNA molecules should have any bearing on how we view the moral status of the embryo.

A third point is that "the program" is not necessarily fixed at conception. For a period of two weeks, it is possible for the developing embryo to break apart into two, three, or even (extremely rarely) four separate fragments that can each develop into a different human being. From where and when do these extra lives arise?

A fourth point is that genetic constitution alone does not define a person. Identical twins may have the same genetic constitution but they are clearly different human beings.

A fifth point is a semantic one. If the word *natural* is used to mean the most likely outcome nature will take in the absence of any external interference (and I know of no other way to define this word in this context), then the *natural* destiny for a human embryo is death. The normal reproductive biology of human beings is such that 75 percent of all *naturally* fertilized eggs will succumb to death *naturally* before the nine-month period of gestation is completed. It is the odd egg only that develops into a live-born baby.

But none of these arguments, or any others, will serve to change the minds of those who believe that human life begins at conception. I can say this because the real reason that people who have thought about the embryo can maintain this point of view is a religious or spiritual one, not a scientific one. Most people do not want to admit that their views of the world are based on spiritual beliefs because in an advanced technological society like ours, with its foundation in science, arguments based on faith alone are not given much credibility. Scientific arguments are required for a cloak of respectability.

Since February 23, 1997, however, science can no longer provide any support for those who hold this point of view. On that date, the cloning of a lamb from an adult cell was revealed, with the assumption that the same technique could be used to clone a human being. The implications of human cloning are staggering in many ways, but one fact, in particular, is relevant to the current discussion: when a new embryo is formed by cloning, *there is no conception.* The new embryo is formed by providing the complete genetic material of an adult cell—like one scraped off your skin—with just the cytoplasm from an *unfertilized* egg. No conception takes place. Actually a conception event did take place a generation earlier, when the person who donated the cell for cloning was conceived. But the new child who will be born from that cell will not have been conceived anew.

If a human life can begin in the absence of conception, then it is scientifically invalid to say that conception must mark the beginning of each new human life. It's as simple as that.

DOES THE EMBRYO DESERVE RESPECT?

Even if you don't think the embryo is morally equivalent to a human being, you can still claim it deserves *respect*. If you believe that it does, you still might want to limit the ways in which its manipulation should be allowed in the laboratory.

What else do people treat with respect besides other people? Broadly speaking, it's other living things as well as inanimate objects that are perceived as important symbols for something else. In my own laboratory, I use mice as experimental animals, and I abide by rules set forth by the National Institutes of Health (NIH) to avoid inflicting any pain that is greater than a human patient might be asked to experience in an experimental situation. I abide by these rules willingly—in fact, I would never do anything to a mouse that was more painful than a needle prick—because I respect the feelings of my mice and I don't want to see them suffer.

But my colleagues down the hall who work on tiny fruit flies abide by no such NIH rules. And most people do not think about the ants they step on as they walk down the street. There are some people who do have respect for even the tiniest form of animal life, but even these people must eat plant cells to survive. And while eating a garden salad, no sane person worries about the live cells being crushed to death as he chews his food.

We treat these different living things differently because mice seem to have feelings, flies and ants probably don't, and plants definitely do not. And it's the *feelings* of animals that we respect, not simply that they're alive.

Do early embryos have feelings? If you reject the notion of a spirit that enters at the time of fertilization (or embryo creation by cloning), the answer is an unambiguous no. Feelings, of any sort, cannot arise in the absence of a functional nervous system, and there are no nerve cells formed during the first week after fertilization. So this is not a basis on which the embryo might be accorded respect. Is there any other?

There are two given by Robertson. The first comes back to the idea that an embryo has the *potential* to form a human being. It makes no sense, of course, to make unrealized potential alone the basis for respect. A newly ovulated egg and each of the millions of sperm in an ejaculate have the potential to participate in fertilization and become a human being, and yet no one believes these cells deserve respect. So potential alone is not enough. Rather it's potential combined with the notion of *complete* genetic identity. The notion of being able to look into the DNA of the early embryo and see a picture of the child it could turn into, given the chance.

There is no denying this connection. Rather, the question is whether it's important enough to induce respect. I'll present two thought experiments that bear on the issue. The first of these is based on methods for manipulating and diagnosing embryos that will be discussed at length in chapters 7 and 17.

Let's start by saying that conception has taken place naturally inside a woman's fallopian tube, and that the woman plans to proceed with the pregnancy in any case, but wants to know as early as possible whether her child will be afflicted with cystic fibrosis. To accommodate her wishes, you remove the embryo at the two-cell stage, take one cell off as biopsy material, dissolve it in a special solution, and carry out your test. You place the remaining part of the embryo (one cell) back into the mother, and it develops into a baby that is born nine months later.

Is this protocol ethically suspect? Based on the intention of the woman to go through with the pregnancy no matter what the outcome, the Vatican has said that prenatal diagnosis is morally permissible. A single embryo was conceived through sexual intercourse, and a single child was born nine months later. Since the diagnostic procedure had no effect on the development or health of the child, no disrespect was committed, right?

The answer is not that simple if you respect the potential of the embryo. Because as soon as you removed one cell from the two-cell embryo, you had two embryos not one. These separate cells could have developed independently into two human beings. By dissolving one cell in solution for diagnosis, you have destroyed a potential life.

Now let's consider a different scenario. Let's say that you separated the two cells in the embryo for the purpose of creating identical twins (rather than genetic diagnosis), as we'll discuss in chapter 7. You now have two separate cells representing two separate embryos with the potential to develop into two different human lives. But you then change your mind. You decide that it's too difficult to raise two children at the same time and you really only want to have one baby. So, you take the two cells and push them back together again to produce the single embryo that you started with.

How does one view the act of bringing together the two cells that you originally separated? You've actually destroyed the *potential* for a second human life, *in the special sense*, without eliminating any life, *in the general sense*. So, have you killed something in the process?

Your answer may provide the foundation for your personal view concerning the respect due to a human embryo with one or two cells. If you do not believe that something has been killed, then, as a matter of logical consistency, you do not respect the potential of early embryonic cells. If you do believe that something has been killed, then "potential" has meaning to you. But what exactly is it that has been killed if not a cell? The only possible answer is a "soul" or "spirit." This concept is what many people are really thinking of when they use the phrase "potential for human life."

With the cloning of Dolly the lamb from an adult ewe cell, and the almost certainty that humans could be cloned in the same fashion, the hidden meaning behind this oft-used phrase is entirely exposed. Now *every cell in your body* has the potential to form a new human life. But no sane person thinks twice about scratching an itch. And if you don't feel bad about killing cells as you scratch yourself, you can't use the idea of "potential" alone as the basis for granting respect to the embryo.

Let us return to the final thing that people treat with respect—inanimate objects. Not just any inanimate objects, but those that have *symbolic meaning* of one sort or another. And even though the embryo is actually alive, this fact does not seem to be relevant to Robertson's second reason for granting it respect: "embryos are potent symbols of human life and deserve some degree of respect on that basis alone."

The problem I see with this rationale is that symbols exist only in the eyes of the beholder—in and of themselves they have no meaning. Consider the American flag, which is perceived by most Americans as a symbol for America, the country. If a foreigner who was unaware of the connection came across an American flag in a foreign land, he would see nothing but a piece of cloth with patterns of stars and stripes. Nothing more could be expected. Clearly, no harm would be done *in his eyes* if he cut up the cloth to make rags for washing his car. And if no one who respected the American flag ever found out about what he had done—an act of desecration *in that person's eyes only*—then no disrespect would have been committed.

Just as one person's flag is another person's rag, so one person's symbol of human life can as easily be another person's clump of cells. In this sense, a symbol is quite different from an "embryonic soul." If you think embryos have souls, then you would logically demand that all people treat all embryos with the respect accorded to human beings. But just because an embryo may be a symbol of human life to you, does not mean that you should expect others to view it in the same way. Instead, you would have to agree that the question of respect is one that each person must decide for himself and herself.

After this long critique of the reasons given for granting the embryo respect, it is time to consider again whether there is a hidden reason lurking in the minds of some (although not all) who hold this position. Perhaps they think that while the embryo is not in possession of a full-blown soul of the kind that you and I have, it does have a "little bit of soul." Perhaps this is the real reason that the Human Embryo Research Panel concluded in their 1994 report to the Director of the National Institutes of Health that "the preimplantation embryo warrants serious moral consideration as a developing form of human life. . . . [even though] it does not have the same moral status as an infant or child."

But a belief in even the smallest bit of embryonic soul implies the existence of something above and beyond complex molecules involved in complex interactions all explained by the laws of physics and chemistry. Indeed, a belief in an embryonic soul, no matter how small or insignificant, can only be maintained by a belief in the vital force. And as we discussed earlier, there is no room for a vital force in any individual living cell, including those present in the human embryo.

4

From Your First Cell to You

If the two-cell human embryo does not represent human life, then what does? When does life, in the general sense, become human life, in the special sense? There is no simple answer to this question. Different people have had different views at different times, and there is nothing that science can add—beyond what's been contributed already—that will provide a solution. It seems unlikely that consensus will ever be reached.

One problem is that development is slow and continuous. Once fertilization is complete, there are no isolated moments along the way where you can point at an embryo or fetus and say that it is substantially different from the way it was a few minutes, or even hours, earlier. It is true that important developmental events, or milestones, can be recognized, but the timing of these events is fuzzy, like the transition from red to orange in a rainbow.

Before one can speak rationally about the origin of a human being, it is critical to gain a feel for the biology of embryonic and fetal development between fertilization and birth. During this nine-month period, the embryo and fetus pass through a series of major developmental stages during which important milestones on the pathway to human life are reached. Through an appreciation for the significance of these milestones, it becomes possible for each of us individually to make an informed decision on the question of the emergence of human life.

DAYS 2–6: CELL DIVISION AND DIFFERENTIATION

At the beginning of the second day after the fusion of egg and sperm, the embryo has two cells. Each of these cells divides, to give four, and each of these divides again to produce a total of eight cells by the middle of the third day. Although the embryo has increased its cell number, it has not done much else. Each of its eight cells, when separated from the others, still has the potential to become an embryo unto itself, and to form a separate human life.

With another round of cell divisions, to produce sixteen cells in total, the first step away from uniformity is initiated. The embryo still looks like a ball, or rather, a microscopic raspberry. But the cells on the outside are able to sense their position relative to the cells on the inside, and in response, they *differentiate* into cells that will eventually become the placenta and other tissues that function to protect the growing fetus.

When biologists use the word *differentiate*, it means *to become different*. When a cell differentiates, it becomes different from the parental cell that gave rise to it. Normally, differentiation causes a reduction in a cell's potential. For example, the placenta-directed cells that have formed as the outer layer of the sixteen-cell embryo have the potential only to produce further cells that will become part of the placenta or other tissues located between the woman and her fetus. These cells have *lost* the potential to end up in the heart, lung, or any other tissue in the developing fetus itself.

Once a cell has undergone differentiation, all of its progeny cells, and their progeny as well, will remain differentiated. But, these cells can still undergo further steps of differentiation with further reductions in potential. For example, by four weeks of age, an embryo contains differentiated cells that have the potential only to produce blood cells. With further cell divisions and further differentiation, cells appear that have the potential only to produce either white blood cells or red blood cells, but not both.

Further rounds of differentiation are required to convert the white cell progenitors into a specific type of white cell that secretes antibodies or another type that gobbles up invading bacteria. At this point, after dozens of rounds of cell division, a state of *terminal differentiation* has been reached. Differentiation and development go hand in hand. Development of a whole organism occurs through the differentiation of the individual cells within it.

Terminally differentiated cells can express highly specialized functions like those just described for white blood cells or others such as the production of body hair or fingernails. Most of the cells in your body are terminally differentiated including all of the microscopic components of complex organs like the lung, liver, kidney, or brain. But there will always be some cells, even in a mature adult, that hold back at an earlier stage of differentiation. These cells are called *stem cells*. They continue to divide to deliver, for example, a new source of skin, blood cells, or other special cell types that must be regenerated constantly in order for you to stay alive.

Molecular biologists now have a sophisticated understanding of what happens inside a cell when it differentiates. Actually, the most important thing is what doesn't happen—a differentiated cell doesn't lose any of its genetic information. Every single somatic cell in your body has a complete set of forty-six chromosomes with all of the DNA that was present in the nuclei of the two-cell embryo that you emerged from. Now, if all cells have the same genetic information, why don't they all look the same and act the same? The answer is that each cell is programmed to use only a small portion of the total information to stay alive and carry out the tasks for which it has been specially designed through evolution. Cells that look and behave differently from one another—as a result of differentiation— are programmed to use different portions of the same total genetic information. And with each differentiative step, the cell program changes in at least a small way.

How a cell differentiates was a fundamental problem in the field of developmental biology for the better part of the twentieth century. But in 1997, it seems fair to say that this problem, in its global sense, has been solved. The solution did not come from any single experiment or laboratory, but rather through the accumulation of many results obtained by hundreds of scientists working around the world. We still don't know all the details, but the general picture is now quite clear. And surprisingly, even to scientists, much of our understanding of human cell differentiation, in particular, and whole body development, in general, is based on experiments performed on simple organisms like yeast and fruit flies. In recognition of the enormous significance of the fly work in particular, my Princeton colleague Eric Wieshaus together with Christianne Nüsslein-Volhard and Ed Lewis won the 1995 Nobel Prize in Physiology and Medicine.

Cells differentiate in response to various kinds of *signals*. Signals can be transmitted between cells that are touching, or between cells that are far away. All of these signals are specialized molecules. Some cells can also receive signals, again in the form of particular molecules, from the external environment.

Through the course of development, and in the adult as well, thousands of signals are being broadcast continuously throughout the body. But like a tiny radio receiver, each cell is tuned to pick out only those signals that are meant for it to receive, and the tuning is based on its state of differentiation. Cells respond to signals by changing the program of genes that they use, which in turn may cause the cell to broadcast its own signals—again, in the form of molecules—to be received by other cells. The developing fetus is like a complex electronic network, except that the signals come in the form of molecules (encoded by genes) rather than electrons. Scientists have yet to discover all of these molecular signals, but the list grows longer every month.

Now when we last left our embryo, it contained sixteen cells and the ones on the outside had begun to differentiate into placental cells (although you wouldn't know it by just looking at them). Cells continue to divide as the embryo continues its journey down the fallopian tube and enters the uterus on about the fifth day after fertilization. Throughout this period, the embryo remains a free-floating independent entity secluded behind its solid zona pellucida coat. Indeed, throughout this period, a woman's body is unable to distinguish between a developing embryo and a decaying, unfertilized egg.

In a certain sense, then, a woman cannot be pregnant during the first week following ovulation. There may be an embryo inside, but it is separate and not dependent upon her for its development. And even if an embryo does exist behind the zona wall, there is still a 50 percent chance that it will pass right through her uterus without her ever knowing it— her next menstrual period will still begin on time. For women who use an Intrauterine Device (IUD) as a form of birth control, the percentage of embryos that pass through jumps up to 100 percent (if the IUD is working properly). These women may have newly formed embryos inside them multiple times each year, even though they never become pregnant.

DAYS 7–13: HATCHING, IMPLANTATION, AND PREGNANCY

Sometime between seven and eight days after the start of fertilization, a large break appears in the zona wall and the embryo comes slithering out in a process that mammalian embryologists actually refer to as *hatching*. Unlike the outside of the zona coat, the outer cells of the embryo itself are rather sticky, and if conditions are suitable, the freed embryo will latch on to the uterine wall. This event marks the start of implantation.

The embryo now invades the uterine wall and establishes connections with the mother's blood supply. For the first time, the presence of the embryo can be detected, both by the mother and by sensitive pregnancy tests. The embryo emits signals that lead to rapid hormonal changes within the mother's body. With an outside source of energy—from the mother, that is—the embryo can begin a period of rapid growth and development. The outer cells of the embryo and the uterine cells of the mother begin to intermingle as they initiate the formation of what will be the placenta.

But even at this stage, the cells in the middle of the embryo have still not undergone differentiation. Each one still has the potential to produce cells that go into every organ in your body as well as nonfetal tissues like the placenta. And the embryo can still break apart into two or more pieces that can each go on to develop a complete fetus and human being. On day thirteen, the curtains close on what IVF practitioners call the pre-embryo stage.

WEEKS 3–5: EMBRYONIC TISSUE DEVELOPMENT

At the beginning of the third week, on the fourteenth or fifteenth day after fertilization, a small number of cells in the middle of the embryo differentiate, for the first time, down the pathway of fetal development. It is only with this differentiation event that it becomes possible to identify *specific* embryonic cells that will definitely be incorporated into the developing fetus. Prior to this point, there was no way of knowing which middle cells would end up in the embryo and which would go into the placenta.

As the embryologist C. R. Austin writes: "Probably, most people unfamiliar with this field would think of the changes as being a continuous line of descent—egg to embryo to fetus to child, each stage being the *full* successor of the one before—but that is not in fact the case. The whole egg certainly becomes the embryo, and the whole fetus becomes the child, but the whole embryo *does not* become the fetus—only a small fraction of the embryo is thus involved, the rest of it continuing as the placenta and other auxiliary structures."

Within the isolated embryonic region that will become the fetus, a line of cells differentiates into a structure called the *primitive streak,* a precursor to the spinal cord and backbone. The appearance of the primitive streak represents a major developmental milestone because it demarcates the point at which twinning can no longer occur. Now if the embryo were to break into two, the separate parts would be unable to complete fetal development. So the fifteen-day-old embryo is committed to the formation of a single human being, or none at all.

Development proceeds very rapidly from this point on. During the fourth week, one can see the beginnings of the gut, liver, and heart. At the end of the fourth week, the heart is beating and primitive blood cells are moving along embryonic veins and arteries. It is also at this stage that the very earliest development of the brain begins. Still, the embryo is less than a quarter of an inch long.

WEEKS 6–14: THE EMBRYO BECOMES A FETUS

Between six and eight weeks after fertilization, the embryo turns into—what appears to be—a miniature human being with arms, legs, hands, feet, fingers, toes, eyes, ears and nose. It is these external humanlike features that cause a shift in terminology from embryo to fetus. By twelve weeks, the inside of the fetus has also become rather humanlike with the appearance of all the major organs. The first trimester of pregnancy is now completed.

Although looks alone can have a powerful effect on how we view something, it is important to understand what is, and what is not, present at this early stage of fetal development. While major organs can be recognized, they have not yet begun to function. Although the cerebral cortex—the eventual seat of human awareness and emotions—has begun to grow, the cells within it are not capable of functioning as nerve cells. They are simply precursors to nerve cells without the ability to send or receive any neurological signals. Further steps of differentiation must occur before they even look like nerves or develop the ability to make synaptic contacts with one another. And in the absence of communication among nerve cells, there cannot be any consciousness. This means that if a fetus is aborted at this stage, it cannot feel any pain.

Ancient philosophers did not have the knowledge of twentieth-century science to help them understand the natural world. Instead, they had to base their ideas on what they could see with their own eyes. And in the eyes of Aristotle, in the fourth century B.C., a fetus looked pretty much like a human being while an embryo did not. So, Aristotle proposed that the human conceptus develops through a series of stages equivalent to the major evolutionary stages of life on earth (although he obviously didn't use the word *evolutionary* since Darwin was not to be born for another two thousand years). The earliest embryo was considered to be vegetative, the later embryo was animal-like, and finally, humanness was achieved at the beginning of the fetal stage. In Aristotle's mind, there was a difference between boys and girls: a boy fetus attained humanness at six weeks, while a girl fetus did not until thirteen weeks after fertilization.

In the thirteenth century, the writings of Aristotle were rediscovered and embraced by the Catholic theologian and philosopher Thomas Aquinas. But Aquinas went beyond Aristotle to argue that the emergence of humanness coincided with the point at which a fetus became "ensouled." Aquinas wrote that at the appropriate time (six weeks for boys and thirteen weeks for girls), God looked down upon the fetus and decided whether it was suitably disposed for ensoulment. If it was, the fetus received a soul and completed the course of development until birth. From the fourteenth century until 1869, this was the official doctrine of the Roman Catholic Church. Before the time of ensoulment, embryos were considered to be a part of the mother's body, and their death or disposal was of no concern.

WEEKS 18–22: THE FETUS KICKS

Between eighteen and twenty-two weeks after conception, the pregnant mother usually feels the movement of the fetus within her for the first time. This moment is traditionally referred to as "quickening." But, while quickening may be an important milestone for a pregnant woman, it is not a milestone of any kind for the fetus. In fact the fetus has been moving around since before the tenth week of its development. It is only when it has grown sufficiently large, though, that its movements can be perceived outside the womb. Typically, a woman will feel fetal movements much earlier in a second pregnancy than in a first, because she is attuned to the sensations.

Again, looks and feelings can be deceptive. Fetal limb movements during this period are not caused by conscious decisions made by the fetal brain. They are simply a consequence of random electrical stimulation of muscle tissue. We can make such a statement with confidence because we know that the seat of consciousness—the cerebral cortex—does not yet have interconnected neurons required for any kind of functionality, even in a primitive way.

In the absence of late twentieth-century scientific knowledge, the movement of the fetus would certainly seem to be highly significant, and the common law in many countries recognizes quickening as the boundary that distinguishes the emergence of a new human life.

Even with our current understanding, it is hard to escape the thought that the kicking fetus inside a woman is a willful little baby trying to escape its confinement. Such thoughts, however, lie in the realm of emotion, not rationality.

WEEKS 24–26: THE FETUS BECOMES VIABLE; THE BRAIN BECOMES WIRED

Two independent milestones occur between the twenty-fourth and twenty-sixth weeks after conception. The first is viability. It is during this period that the fetus develops the ability to survive outside the womb. Survival becomes possible as the fetal lungs begin to function for the first time. Even with the best neonatal technology available, we cannot push the point of viability back any further simply because a younger fetus cannot breathe. A number of other organs are not yet fully functional either, and even in the absence of the lung problem, there would be other roadblocks to earlier survival.

What about the future? Will medical science be able to overcome the problems inherent in the survival of the early fetus? And beyond that, is there any chance that an artificial womb can be developed to take the place of the biological womb throughout the entire pregnancy, as described in Huxley's *Brave New World*?

Fortunately or unfortunately, depending on your point of view, this is an extremely difficult technical problem. From implantation until the twenty-fifth week of pregnancy, the conceptus lies in intimate contact with its mother's body. The fetus receives all of its nutrients from the mother and passes all of its waste products to her for excretion; this is well-known. What is not understood are all of the molecular signals that must be flying back and forth throughout development to fine-tune the process by which fetus and mother respond to each other. To create an artificial womb, one would not only have to understand what each of these signals means, one would also have to program the "a-womb" to respond to each signal in an appropriate biochemical manner.

One should never underestimate the power of future technology. It is certainly possible, if not likely, that an artificial womb will be developed over the coming centuries so long as research is allowed to proceed. But in comparison to most of the other reprogenetic technologies that will be discussed in this book, the a-womb seems to be on the more distant horizon.

The second critical milestone that occurs between the twenty-fourth and twenty-sixth weeks is the emergence of a functional cerebral cortex, and with it, the potential for human consciousness. This process begins at twenty-five weeks and is described in dramatic terms by Morowitz and Trefil in their book, *The Facts of Life*:

> *Most brain cells are produced early in the pregnancy, migrate to their final position, and mature into their final form. During this period, a few synapses form, but there is no large-scale wiring up. Then when most of the cells are in place and all is in readiness, synapses start forming in earnest. It is this burst of synapse formation that we call the birth of the cerebral cortex. It marks the period during which the*

brain is transformed from a collection of individual cells into a connected machine capable of carrying out human thought.

Based on the developmental and anatomical evidence, Morowitz and Trefil argue that human life—in a sense somewhat different from that used in this book—begins sometime after twenty-five weeks of development. Interestingly, as these authors note, by sheer coincidence, "humanness and the ability to survive outside the womb develop at the same time."

Strictly speaking, it seems unlikely that the newly emergent cerebral cortex has the ability to carry out conscious thought. But, there is no simple test, or even definition, of what conscious thought is. Although both synapse formation and organized cerebral electrical activity (measured as recognizable EEG brain-wave patterns) begin at twenty-five weeks, they both continue to change and mature until a child reaches the age of ten! Obviously, consciousness emerges very early on, but when exactly that is, nobody knows.

BIRTH AND BEYOND

The average time from conception to birth is 270 days, but as we saw, survival is possible as early as 25 weeks with sophisticated neonatal care, and some pregnancies can last as long as 40 weeks. Thus, birth by itself does not mark a specific point in development. On the other hand, by 35 weeks, the fetus has developed to the stage where it can survive in the outside world on nothing but its mother's milk or a synthetic substitute.

In most societies, a newborn baby is considered to be "the most precious thing in the world." We look upon her or him as a complete human being whom we treat with at least the same respect and care that we would give to other human beings. As a starting point for discussion, most readers of this book would agree that they did *not* exist before fertilization and that they *did* exist at the time of birth.

We are still left with the question that we have considered throughout this chapter: when between these two points did you begin your life— your *human life*? When your father's sperm penetrated the zona coat around your mother's egg? When a single sperm cell was gulped into the egg's cytoplasm? When the sperm cell stopped swimming, lost its tail, and expanded its nucleus? When the egg eliminated half of your mother's DNA so that the genetic material present was finally equivalent to that currently in each of your cells? When the genetic material from your mother and father first moved closer together in each of the two cells of the two-cell embryo? When the first definitive cells destined to end up in your body appeared fourteen days later? When twinning was no longer possible, a short while after that? When the most primitive portion of your brain first appeared between three and four weeks? When the cerebral cortex emerged between six and eight weeks? When your cerebral cortex became wired at twenty-five weeks? Or when you left your mother's body and began to breathe on your own sometime between twenty-five and forty weeks after fertilization?

There are a few who would argue that the real beginning did not occur between fertilization and birth. If one accepts the definition of human life that I've presented, it is possible to conclude that you didn't exist as a unique human being until you became fully conscious of the world around you, or fully self-reflective, months or perhaps years after you were born.

On the other hand, it is also possible to go back before fertilization and consider the significance of what existed then. While the information that would eventually be used to mold your body was not yet in one place, all of it did exist in a discrete form somewhere. Furthermore, the actual egg cell that developed into you was produced by your mother's body when *she* was still a fetus.

This fascinating little fact intrigues Brigid Hogan, the co-chair of the National Institutes of Health committee that was charged with recommending policies for embryo experimentation in the United States. When I asked her for her views on the question of life's beginnings, she said: "I like the idea that a long time ago, in Port Elizabeth, South Africa, a young pregnant Englishwoman had inside her body not only her daughter but the egg that gave rise to her granddaughter and that the genetic recombination that contributed to me started then." Of course, if one continues to travel backward in time to look for a beginning, there may be no stopping until the first cell is reached, 3.5 billion years ago.

What science tells us is that there is no single moment that marks your beginning. No single moment that can be isolated away from so many other important moments and that we can all agree upon. Instead, a scientist will tell you that you emerged slowly over time from the genetic information and molecules that made up your developing body. And what

I will describe to you in the chapters ahead are the ways in which scientists have learned how to manipulate that information and those molecules so that in the near future, we as a species will have the power to control the very nature of the human lives that emerge.

CREATING LIFE

All things were made by him; and
without him was not anything made that
was made. In him was life, and the life
was the light of men.

—The Gospel according to John 1:3–4

5

Babies Without Sex

During a seminar on reprogenetics that I taught to freshmen at Princeton University in 1995, I revealed a little secret that they all found rather amusing. What I told them is that every one of their ancestors—both parents, all four grandparents, all eight great-grandparents, sixteen great-great-grandparents, and continuing on back as far as they could imagine—all of them, without exception, engaged in sexual intercourse with a member of the opposite sex. Young people often find it hard to believe that their parents may have engaged in sexual intercourse, but their grandparents and great-grandparents as well? Shocking.

And yet, at some point in the future, there is sure to come a time when I will find a student in one of my seminars for whom this will not hold true. For 600 million years after its invention, sexual intercourse is no longer a prerequisite to reproduction. Consider the following news stories.

On December 7, 1987, the *Washington Post* reported:

Recently, Episcopal priest Lesley Northrup has attracted the world's attention by becoming a single mother through artificial insemination.

She has argued that the technique avoids the church's ban on sex outside of marriage, and her bishop has supported that position.

"I have done nothing illegal or immoral in having this child," Northrup said. "I cannot think what the offense might be. Adultery? None of the parties was married. Extramarital sex? No sexual act occurred."

On July 19, 1994, the *Sunday Telegraph*, a British newspaper, announced: "In Buckinghamshire a woman who says that she never 'fancied' any man enough to go to bed with him has a son by a 'virgin' birth."

And on January 19, 1995, the *Daily Mail* (also in Britain) reported:

Two jobless lesbians have become the parents of a baby girl following a successful DIY (artificial insemination) pregnancy. Both claim they are virgins and the child's mother says she conceived after inseminating herself with sperm donated by a gay male friend. . . . The space for the father's name on the birth certificate will remain blank. Natalie [Wilson] said they had longed for a baby but thought it would be impossible since neither was prepared to have sex with a man to become pregnant. "Now against all the odds, that dream has come true," she said.

Other reports of virgin births have come to light (especially in the British tabloids), and for each reported case, there are probably hundreds, perhaps thousands more that go unreported. Nearly all of them have been initiated with the use of artificial insemination by donor (also known as AID or DI), which is the least technical of the reproductive technologies. DI can be performed by a woman on herself with just a turkey baster containing donated semen. And even the turkey baster may not be strictly required. In the novel *Galápagos*, Kurt Vonnegut describes acts of artificial insemination that are initiated with semen inserted into the vagina with fingers alone. In Vonnegut's vision, these children conceived *in the absence of sin* are the Adams and Eves for the entire future of human life on earth.

Has something immoral occurred with the sundering of the link between sex and reproduction? Some think so. The *Washington Post* on March 12, 1991 reported:

Several members of [the British] Parliament from the ruling Conserva-
tive Party . . . called for legislation that would ban artificial insemina-
tion of virgins . . . "It is difficult to imagine a more irresponsible act
than to assist a woman to have a child in this highly unnatural way,"
complained Dame Jill Knight, a Conservative lawmaker. An embar-
rassed government said a ban on virgin births would be unworkable.
That was not quite good enough for some Tory legislators. "I find it
personally abhorrent," said Jerry Hayes, chairman of the party's health
committee. "One virgin birth for eternity is enough."

What is it that so upsets these British legislators? It is not just the fact
that children are being born to unwed mothers; that's old news. And it's
not just that lesbians are having children, for the anger is directed at
nonlesbians like Lesley Northrup as well. It is, instead, a sense that there
is something *highly unnatural* about an act of reproduction from which
men have been excluded.

In speaking about those who were most angry at what she had done,
Northrup said, "I am not a radical feminist, you know, but it's interesting.
They were all men who complained that they had been taken out of the
process. The fact that one can start a family, build family life, without a
male being in it seems to be threatening."

Suddenly, it seems to some, women have the power to gain complete
control over their reproductive destinies. For the present, men are still
required to contribute their sperm, but with the advent of human cloning,
they won't be needed at all. And if men are left out, does that mean that
God is somehow left out of the process as well? Will Woman alone—with
a capital W—become the creator of all new human life?

The idea that God has been excluded is at the root of the Vatican's
condemnation of IVF and most other reproductive technologies. For it is
the creation of babies in the absence of sexual intercourse, rather than the
technology itself, that is considered immoral. In fact, the Vatican seems
willing to accept a variation on the IVF protocol that begins with the
infertile couple undergoing intercourse while the husband wears a special
condom in which small holes have been cut purposely to abide by the
Church's edict against contraception. After intercourse is completed, sperm
are retrieved from the special condom and placed in a laboratory dish
next to the eggs retrieved earlier from the wife's ovary. Then the whole
mixture is quickly inserted back into the wife's fallopian tube so that

fertilization can occur on the inside rather than the outside. This protocol has been given the acronym GIFT, not only to distinguish it from the morally suspect IVF technology, but also to provide the image of a "gift of life" that comes directly from God.

The Vatican seems willing to accept this ruse because it can identify a specific act of sexual intercourse connected to a specific act of fertilization within the *natural* environment of the wife's reproductive tract, even though an act of high technology has intervened between the two. Since every sex act between married people is "willed by God" (according to the Church's teachings), it follows that every child that emerges from such an act is God's creation. In contrast, when women conspire to bring about fetal development in the absence of sex, not only do they create children in the absence of "God's will," they actually steal the creator's scepter for themselves.

I find it amusing that while mostly older, conservative men see a horrible future in which they and their sons are left out of the reproductive process, a few women on the other side of the political spectrum fear a future that is exactly the opposite. Gena Corea, a feminist scholar who has written extensively on reproductive issues, believes that "reproductive technologies . . . are transforming the experience of motherhood and placing it under the control of men . . . Reproductive technologists now aim to bring forth life through 'art' rather than nature and enable a man to be not only the father but also the mother of his child."

No matter who ends up with the power—and I suspect it'll be equally distributed between men and women—the British Conservatives and Gena Corea are both right. Our approach to babymaking is in for a big change. Today, sexless acts of reproduction represent only a tiny fraction of the babies born in industrialized countries. But every year, more and more babies will be conceived in this way. And if technologies like cloning, embryo screening, and genetic engineering come into widespread use during future centuries or millennia, sexless reproduction could become the norm for those people who have the money to pay for it.

Except for artificial insemination—which is so low-tech that it hardly deserves designation as a technology—all of the many other reprogenetic possibilities we will discuss build upon the foundation established by in vitro fertilization.

6

In Vitro Fertilization and the Dawn of a New Age

IN THE BEGINNING

A singular moment in human evolution occurred on July 25, 1978, with the birth of a baby girl to Lesley and John Brown in the Oldham and General District Hospital in the town of Oldham, England. Mrs. Brown did indeed have a lovely daughter. She weighed 5 pounds 12 ounces and was named Louise Joy. Louise was extraordinary because she was the first human being conceived outside her mother's womb.

Nine months earlier, a single egg had been removed from Lesley Brown's ovary and placed into a small plastic dish by Patrick Steptoe. Sperm obtained from John Brown were added to the same droplet of culture fluid and the dish was placed under the microscope where Steptoe's colleague Robert Edwards watched as fertilization took place. The fertilized egg was allowed to divide three times and was then placed into Mrs. Brown's uterus. Nine months later, Louise Brown was born.

This birth represented the culmination of more than a decade of work on human eggs and embryos by Steptoe and Edwards, who should be recognized as the founders of the new age of reprogenetics. The significance of their technical feat cannot be overemphasized. IVF—the term now used to describe the entire process from egg and sperm collection to embryo placement in the uterus—was developed originally for the purpose of curing one type of infertility. But what IVF does inherently, as well, is provide access to the egg and embryo. And with this access, it becomes possible to observe and modify the embryo and its genetic material before a pregnancy is initiated.

Robert Edwards—the original figure behind the development of IVF— did not have a medical degree. His graduate training was at the Institute of Animal Genetics at Edinburgh University, and his Ph.D. thesis work performed during the early 1960s was on mouse embryos, not those of humans. As a graduate student, Edwards took advantage of techniques perfected by other mouse embryologists for the fertilization of eggs in culture dishes and the transfer of embryos back into female mice where it was found that development proceeded normally to the birth of healthy animals. It was his experience with mouse embryos that convinced Edwards that IVF could be made to work in humans as well.

So the push for human IVF did not come from within the medical establishment. It came instead from a basic researcher who understood and appreciated the striking similarity in biology that exists between humans and all other mammals. This point underlies much of the rationale I will use for predicting future advances in reprogenetics.

In 1968, Edwards persuaded Steptoe, a gynecological specialist, to join him in his quest, and together they spent the next ten years seeking to replicate the mouse IVF protocol in human subjects. The problem was not in getting fertilization to take place in a dish; that task was accomplished very early on. The problem was in working out the conditions required for the embryo to implant into the mother's uterus. Figuring out all of the details is much more difficult to do when there's no opportunity to experiment, and obviously, the introduction of every embryo into a woman's uterus was made with the hope of success and not just to collect experimental data.

AT THE PRESENT TIME

From a single birth in a British clinic, the use of IVF as a method of reproduction has exploded. In 1985, the first year that a survey of U.S. clinics was performed, 337 births were reported. In 1990, the number of American IVF births jumped to 2,345, and in 1993, it rose to 6,870. In 1990, there were 180 U.S. clinics offering IVF services; by 1993 that number had jumped to 267.

And the United States is far from alone. IVF programs of comparable size are churning out babies in Australia, France, Belgium, Holland, and, of course, Britain. Perhaps surprisingly, IVF is not just confined to wealthy countries of the West. By 1994, more than thirty-eight countries had established IVF programs, including Malaysia, Pakistan, Thailand, Egypt, Venezuela, and Turkey, each of which has reported more than one hundred births. The total number of IVF babies born by the end of 1994 was estimated to be 150,000 and most of these children are less than four years old. It is now likely that you or one of your friends knows someone who has had a child in this way. If IVF services continue to expand at a comparable rate, by the year 2005, there could be more than 500,000 IVF babies born annually in the United States alone, and millions more in other countries.

These numbers are astounding because IVF is not a technology that people can master in their spare time. It requires highly skilled physicians and reproductive biologists. Physicians must go through medical school and a long residency program in obstetrics and gynecology followed by a period of specialty training at an IVF clinic. And training alone is not enough to ensure success; some have the "right stuff" for performing feats of microsurgery and micromanipulation, while others never will.

Even with the large hurdles that must be overcome, it is easy to see how the number of IVF providers continues to expand. First, there is an enormous pent-up demand for IVF services for reasons I will discuss below. Second, the profits to be made are large. At a typical IVF clinic, a couple may spend between $44,000 and $200,000 to achieve a single pregnancy. Unlike the computer industry where equipment today can be purchased for a fraction of what a comparable machine cost a few years earlier, the cost of IVF services is unlikely to decline in the future, since most of it is accounted for by labor. And the labor we're talking about here is that of the highly skilled medical professional who will always insist on a large fee for his or her highly specialized services. In 1992, reproductive specialists on the staffs of hospitals, HMOs, and group medical practices were paid an average annual salary of $259,750, which was more than any other medical specialty. And IVF practitioners who have a stake in their own private clinics can make much more money than that.

So, why are there so many people who are willing to pay such large amounts of money, often entirely out of their own pockets, for these services? The answer lies in the powerful desire to have *a child of one's own.*

THE DESIRE TO HAVE A CHILD

The desire to have and raise a child is such a powerful instinctive force that many people who experience it have a hard time explaining where it comes from. But the source is readily apparent to those familiar with Dobzhansky's famous quote, "nothing in biology makes sense except in the light of evolution." In this light, the origin of the desire is easy to see. It emerges directly from one of the guiding principles of evolution: genes that program individuals to do a better job at reproducing themselves will be passed down with increased frequency from one generation to the next, and will eventually spread widely throughout a population.

Which of the 100,000 genes that we carry increase the efficiency of reproduction? Actually, they all do, otherwise they wouldn't be present within our genetic material. But most operate in ways that simply help our bodies survive long enough to reproduce, rather than in the process of reproduction itself. Of the genes that are involved directly in reproduction, most affect the physiological processes of sperm and egg production. Infertility is often caused by the dysfunction of one or more of these genes.

In addition to genes that control the development, physiology, and structure of an animal are those that control behaviors and emotions that benefit the process of reproduction. There are genes that program a male bird to court a female and others that program the female to choose her mate wisely. There are genes that program a male dog to risk his life to copulate with a female in heat. And there are genes that control the development of the human brain so that it is predisposed to express behaviors and emotions that are beneficial to the reproductive success of the body it inhabits.

There are many examples of instinctual behaviors—beneficial to reproduction—that we as humans share, to one degree or another, with certain animals. One of these is an innate fear of snakes, which helps to keep us alive before and during our reproductive years. A second is the desire to engage in sexual intercourse. In addition, there are those that are uniquely human, and one critical human instinct is the abstract desire to have children.

It is easy to imagine how such a desire might have evolved in our ancestors. It probably began with the ability to generate and process abstract thoughts and make logical connections between events that occurred far apart from each other in time and place. The fossil evidence suggests that our ancestors gained this intellectual capacity between one and three million years ago, during a period when the cerebral cortex underwent a large expansion in size. And what the increased intellectual capacity provided (as a byproduct) was the ability to make connections between sex, pregnancy, and babies. Once these connections were made, the stage could have been set for the evolution of the desire to have children.

People whose genes programmed them with this reproductive desire (separate from a sexual desire) would be more likely to engage in activities that promoted successful pregnancy, childbirth, and parenting. Their children, in turn, would inherit the same genes and do the same for their children, and so on through each generation. Ultimately, the emotional desire to have children would spread throughout the entire species.

Of course, most of us know people who are childless by choice. How does biology explain this? The explanation comes from the single attribute that uniquely defines us as human beings. We alone—among all animal species—have evolved the intellectual capacity to comprehend and, at times, counteract the natural predispositions provided to us by our genes. And under certain circumstances of environment, culture, or intellect, reproductive desires can be rejected in favor of other desires centered more on the self, on other human beings, or other life goals.

For the vast majority of people, though, the desire to have children is so powerful that it outshines everything else they might possibly want to do during their lives. And the inability to fulfill the desire may be accompanied by a degree of pain and grief equivalent to that felt upon the death of a loved one. Unfortunately, 9 to 15 percent of all married couples are infertile. In the United States alone, there are more than two million couples right now who want to conceive and are unable to do so.

INFERTILITY CAUSES AND CURES

There are many causes of infertility. A man may not be able to produce sperm, he may produce too few sperm, or the sperm that he produces may be ineffective at completing the process of fertilization. Similarly, a woman may not be able to ovulate, she may produce eggs that are resistant to fertilization, or the zona coats around fertilized eggs and embryos may be resistant to the hatching that must occur before implantation can take place. There may be other problems as well: A woman's fallopian tubes may be blocked so that eggs can't pass through, her reproductive tract might be chemically hostile to sperm that enter, or her immune system may destroy sperm as if they were foreign invaders. Any one of these problems can stop the reproductive process cold, so it's not surprising that infertility is so pervasive.

Steptoe and Edwards developed IVF as a method to treat only a small fraction of infertile couples, the 15 percent whose infertility was caused by a single problem—blocked fallopian tubes. With IVF, the fallopian tube can be bypassed entirely; an unfertilized egg is taken from the ovary and the product of fertilization is placed back into the uterus.

Today, IVF can be used as a starting point to treat nearly every form of infertility that is not responsive to treatment by less invasive means. A reproductive tract or immune system that is hostile to sperm is no longer relevant. If a woman doesn't ovulate naturally, hormones can be used to induce ovulation. If a man's sperm concentration is too low, it can be increased in a test tube. If sperm can't make it through the zona themselves, they can be injected into the space between the zona and the egg. If the zona around the embryo is resistant to hatching, it can be manually disrupted in the laboratory before the embryo is placed into the uterus for implantation.

And if for any reason sperm and egg refuse to fuse no matter how well they are coaxed together, that's no longer a problem either. For a single live sperm can be picked up in a tiny glass needle and injected directly into the egg cytoplasm, avoiding the process of fusion altogether. This protocol has been named Intra-Cytoplasmic Sperm Injection, and is given the acronym ICSI. Although it was only developed in 1992, four years later, it was already used in a third or more of all IVF procedures performed at many large clinics in the United States and elsewhere. Up to 80 percent of the fertilized eggs that are created this way can proceed normally to at least the two-cell stage of development. And the overall success rate—measured in terms of development from an embryo to a live-born baby—is no different from that achieved with traditional IVF.

It is interesting to consider a philosophical implication of ICSI for those who believe that human beings come into existence at the moment of fertilization. Although fertilization occurs over an extended time, the moment that most people think of when they use this phrase is when the sperm and egg fuse together. With ICSI, there is no sperm-egg fusion, but it can be argued that the moment of fertilization—and the creation of a human being—now occurs with the manual injection of the sperm into the egg cytoplasm. A central tenet of this philosophy is still the notion that the moment of fertilization represents a point of no return, when two cells cease to exist and in their place, there is just one. As the Vatican writes, "from the first instant, the program is fixed . . ."

But during a brief time after injection, the sperm is still alive as an independent cell—swimming in the egg cytoplasm—in the same sense that it was before injection. Furthermore, the rapid-response events induced by sperm-egg fusion (such as the barrier erected to prevent additional sperm from entering the egg) simply fail to occur with ICSI. This means that it's possible to go back into the egg that we just fertilized and pull the sperm cell out; with this action, we could reverse the process and return to where we were before—with two living cells instead of one and no embryo. Thus, the real point of no return is no longer coincident with the point of fertilization. Of course, with cloning, a human being could be born in the absence of any fertilization event at all.

With the development of ICSI, the concentration and quality of sperm produced by a man are no longer an impediment to fertilization. But what if the man produces no sperm at all? Wouldn't this be an insurmountable obstacle to reproduction?

Not anymore.

The sperm cells produced by a fertile man are formed through the differentiation of round tailless cells called spermatids. Spermatids are testicular cells formed immediately after the genetic material has been reduced by 50 percent as I described in chapter 3. Most men who fail to produce sperm *do* carry less differentiated spermatids in their testes. And IVF practitioners have perfected ways for recovering these spermatids from the testes of infertile men and then plucking the nuclei—containing the genetic material—out of these cells. A single naked nucleus is then injected into the egg cytoplasm to initiate fertilization. The process is called Round Spermatid Nucleus Injection or ROSNI.

But what about those men who fail even to make round spermatids? These men have no cells that have undergone a reduction in genetic material. So these men have no source of nuclei that could be used for injection into the egg. Incredibly, it now seems that even this severe defect will soon no longer be an insurmountable obstacle to reproduction.

Ralph Brinster, at the University of Pennsylvania Veterinary School, has developed methods for taking the most immature stem cells in the testes, called spermatogonia, and placing them into a "foster testes" where they can differentiate normally into fully active sperm. It is even possible to take cells from one species and let them differentiate in the testes of another. The implication of Brinster's results is that immature testicular cells could be taken from a man whose own testes did not support differentiation, and placed in a pig's or bull's testes, for example, that would now produce human sperm for use in IVF. We will come back to the bizarre implications of this practice in chapter 14.

REPROGENETICS: BEYOND IVF AND INTO THE FUTURE

Earlier I said that the birth of the first IVF baby represented a singular moment in human evolution. Yet medical science in the twentieth century has had enormous success developing cures for many once-fatal illnesses. Why should a cure for infertility—and an imperfect one at that—be singled out as more important than all of the hundreds of other medical advances that have occurred during our lives? Aren't cures for diseases that used to kill or lame children, in particular, more significant to our society?

I don't think a cure for infertility should be placed on a higher pedestal than the development of a polio vaccine or cures for childhood cancers. But this isn't what I had in mind when I used the phrase "singular moment." Rather, it was the conviction that although IVF was developed as a means for treating infertility, it will now serve as a stepping stone to many reprogenetic possibilities that go far beyond its original purpose. And because some of these possibilities may very well "change the nature of our species" as the editors of *Nature* put it, the development of IVF marks the point in history when human beings gained the power to seize control of their own evolutionary destiny. In a very literal sense, IVF allows us to hold the future of our species in our own hands.

The possibilities that open up with the successful use of IVF can be grouped into three broad categories. The manipulation of the embryo at the cellular level is the focus of Part II of this book, which you are now reading, as well as Part III. The second category involves the ability to move embryos from one maternal venue to another with implications for the meaning of motherhood, which is the focus of Part IV. Finally, it is by bringing the embryo out of the darkness of the womb and into the light of day that IVF provides access to the genetic material within. It is through the ability to read and alter genetic material inside the embryo that the full force of IVF will be felt, which is the focus of Part V.

WILL IT HAPPEN?

Before I describe the reprogenetic objectives made possible by IVF, it is important to consider whether people would actually be willing to sever the link between sexual intercourse and babies in an attempt to achieve some sort of reproductive goal, and whether they would be able to find professionals willing to work with them on the task. It depends, of course, on what the goal is. There's a big difference between curing infertility, on the one hand, and trying to make sure that your child inherits your curly hair on the other. More than 75 percent of Americans now feel that IVF is an acceptable solution to infertility, while many fewer accept its use for purely cosmetic reasons. But there are many reprogenetic goals that lie between these two extremes. Where will people draw the line?

No matter where it is drawn today, it will almost certainly be drawn to include more reprogenetic possibilities in the coming years, and more still in the years after that. This is because breakthrough technologies are always viewed as alien when they first appear—many people are instinctively opposed to things they are not accustomed to. But as the physicians Sophia J. Kleegman and Sherwin A. Kaufman observed in 1966: "Any change in custom or practice in this emotionally charged area [of assisted reproduction] has always elicited a response from established custom and law of horrified negation at first; then negation without horror; then slow and gradual curiosity, study, evaluation, and finally a very slow but steady acceptance."

The public's opinion of IVF has evolved in this very way. When news of its development by Steptoe and Edwards reached the media during the 1970s, there were editorials calling for the abandonment of all further research on "test tube babies." And when the first IVF baby was born, most Americans found the notion so bizarre that they wouldn't think about using it themselves. Over the ensuing decade, however, IVF has been transformed from an alien concept to a broadly accepted medical approach for treating infertility.

Let's consider the arguments that can be made against the possibility that IVF will be used for purposes other than the alleviation of infertility. The first argument is that people will not be willing to subject themselves to an alien technology that separates sex from reproduction just for the purpose of providing their child with some advantage that she might not otherwise have. Either ethical or emotional concerns, or both, could be at the root of this unwillingness.

The second argument concerns cost. Even if people had no objections to using the technology per se, they might not be willing to spend $30,000 or more for this purpose.

The third argument is that even if people were willing to pay, they wouldn't be able to find clinics that were willing to provide the *nonessential* reprogenetic services that they desired. This could be for two reasons. First, the technical expertise itself might not be available. Second, those with the technical expertise might have ethical objections to using it in this manner.

There is no doubt that in Western societies today, many people have a strong gut reaction against the use of reprogenetic technologies for non-medical purposes. I observed this gut reaction when I asked a class of about one hundred seniors in a 1996 "Biotechnology and Society" course at Princeton whether they would ever consider the use of genetic engineering on their own children-to-be *for any reason*. More than 90 percent said no. But when I presented a hypothetical scenario in which genetic engineering might be used to provide absolute protection against AIDS, and posed the question again, half changed their minds. In a matter of minutes, they switched from rejecting a reprogenetic technology to accepting it when presented with a specific example.

What about the cost? Would $30,000 be too much to pay to ensure that a child would be born healthier or wiser, in some way, and better able to compete in the world? In fact, it is not uncommon for American parents to spend three times more than $30,000 over a four-year period to provide a child with a college education. And what is the point of this expenditure? It's to increase the chances that their child will become wiser, in some way, and better able to achieve success and happiness. If parents are willing to spend this money—with no guarantee of a return on their investment—after birth, why not before?

Parents might be *willing* to spend this money, you might say, but only the wealthy will be able to afford it. This notion is belied by the entry of so many middle-class couples into current IVF programs. In a case that we'll discuss in chapter 7, a Tennessee couple with a joint annual income of just $37,000 was able to come up with the money required for seven separate IVF attempts at pregnancy over a four-year period.

Finally, there's the question of whether there will be clinics that are willing to provide these nonessential services. In this regard, there can be no doubt. IVF practitioners are expanding so rapidly that they are bound to reach a point where the pent-up demand from infertile couples is satisfied. And when this point is reached, if not sooner, some will go looking for new customers.

Many practitioners, including those associated with major medical centers, may worry about ethical or political concerns before proceeding. But consider the countries where IVF is being practiced successfully today; consider as well the hundreds of private clinics that operate in the United States; consider the amount of money to be made; and consider that as of August 1997 there are no federal laws that regulate private IVF practitioners in terms of the services they can offer to their clients. If there are people who desire reprogenetic services, there will be others willing to provide them.

You may protest still that IVF practitioners have learned how to play with embryos but don't yet know how to play with genes. This may be true today, but genetic technology is even more widespread than IVF technology, and it can be learned with much less schooling. Every year, my colleagues and I send Princeton students with nothing more than a bachelor's degree into the world with the knowledge and training required to perform the genetic "miracles" I will discuss in this book. And every year, thousands of other young people complete their graduate studies in molecular biology and genetics programs at hundreds of universities. It seems inevitable that some of them will team up with IVF practitioners and reproductive scientists to provide reprogenetic services to the world.

7

Frozen Life

COMING IN FROM THE COLD

"Suspended animation" is the popular term used to describe what happens when a living thing is frozen with the assumption that it will be thawed later to get on with its life. Scientists use the word *cryopreservation* to describe the same process. The *cryo* prefix comes from the Greek word *kruos*, which means icy cold, although the actual temperature used for cryopreservation (–196° Celsius or –320° Fahrenheit) is far below the freezing point of water.

Freezing living things is easy, of course, but getting them to become reanimated upon thawing is more difficult. Obviously, the success of cryopreservation must be judged—after the fact—by how well things come back to life. Cryopreservation was first performed with success on various types of cells during the 1940s. In 1950, it was used successfully on bull sperm, and in 1953, it was shown that human sperm could also be frozen, stored for long periods, and then thawed with the ability to come out swimming, ready and able to fertilize an egg through artificial insemination. Today, sperm from both bulls and men are routinely stored frozen in sperm banks around the world.

The first embryos to be successfully frozen and thawed were those of the mouse, in 1971. This success was followed within a few years by the successful cryopreservation of embryos from rabbits, sheep, goats, and cows. And, on March 28, 1984, Zoe Leyland became the first human child to be born from a frozen embryo in Melbourne, Australia. Since then, cryopreservation of human embryos has become a routine practice in all well-established IVF clinics.

Just like the procedure of IVF itself, the successful application of cryopreservation requires attention to all sorts of details, including the speed at which the temperature is lowered during the process of freezing, the speed with which it is raised during thawing, and the physical and chemical environments that surround the embryo while this is all taking place. Researchers started by using conditions that worked well for various animal species, but in order to optimize the chances of successful human pregnancies, they needed to tinker directly with human embryos. With each experiment, one condition or another was slightly modified, and its effect on survivability and developmental potential after thawing was determined. Through this trial-and-error process, the chance of pregnancy with cryopreserved embryos has reached—in some clinics—that obtained with unfrozen embryos.

IS THE FROZEN EMBRYO ALIVE?

This is a problem of semantic as well as philosophical curiosity that we can try to solve by examining the attributes normally ascribed to biolife in its general form. The first consideration is whether the frozen embryo is using energy in the way that living things do. It is not. All of its molecules are at a near standstill, just vibrating slightly but not doing anything else. The animation it once had is now suspended. In this sense, a frozen embryo is no different than a frozen *inanimate* object. In this sense, a frozen embryo is not alive.

Next, we can ask whether the frozen embryo retains the structure and information that is characteristic of living things. Yes, it can. And not only does it retain these attributes, but it may continue to do so for hundreds of years, and even longer, if it is kept at the same temperature. In this sense, the frozen embryo is alive.

A final question that might help us to resolve this stalemate is whether the frozen embryo has the *potential* to become reanimated. Can it live again? Here there is no absolute answer. With the procedures used for cryopreservation today, some embryos are completely dead upon thawing, others are completely alive, and others still have a combination of both living and dead cells. If a sufficient number of cells in these mixed embryos is alive, the embryo as a whole can go on to develop into a live-born child.

The difference between life and death is caused by small differences in the microenvironment experienced not only by each embryo, but by each embryonic cell, during the process of freezing or thawing. It may be impossible to know while an embryo is frozen whether it will have the potential to form a developing embryo again, let alone a live-born child. This ambiguity makes "the potential for life" attribute inadequate for determining the living status of any individual frozen embryo before it is thawed.

One is therefore drawn to the conclusion that the frozen embryo is neither alive, nor dead, but rather in a third, entirely different state.

WHY FREEZE HUMAN EMBRYOS?

Embryo freezing plays an important role in a typical course of IVF treatment conducted at the most technologically advanced IVF clinics. It allows the retrieval and fertilization of up to thirty eggs during a single operation. Three or four can be transferred back into the woman immediately, while the others are frozen away in small groups. If pregnancy is not achieved, a group of embryos can be thawed and transferred at the critical point in her cycle during each subsequent month. Thus, the freezing protocol provides a way for women to avoid the physical stress of repeated hormonal stimulation of ovaries and egg retrieval. Furthermore, since this part of the procedure can account for up to 90 percent of the cost, the financial burden caused by multiple tries at IVF is greatly reduced.

A second medical application of embryo cryopreservation is in special cases where women suffering from certain medical ailments are likely to lose the function of their ovaries or have their eggs destroyed or harmed. This could happen, for example, when a woman undergoes chemotherapy or needs to have her ovaries removed as a treatment for cancer. In these cases, and others, a young woman may not yet have made decisions about her reproductive goals. Egg and embryo freezing can provide her with the option of putting off a decision until some time later.

A third medical use of cryopreservation is in cases where eggs or embryos are donated by one woman to be used by another who is unable to produce her own. It is difficult to synchronize the ovulatory cycles of two women, yet this is essential for successful implantation when embryos are transferred directly from one woman to another. Cryopreservation eliminates this problem. Donated eggs can be fertilized and kept frozen until the appropriate time is reached in the second woman's cycle, when her uterus is most receptive to implantation. I will discuss egg and embryo donation further in chapter 13.

Embryo freezing can also be used for the purpose of genetic diagnosis. As we will discuss in chapter 17, it is possible to remove one or a few cells from individual embryos and then use sophisticated molecular techniques to determine whether particular genes are present or not. If the diagnostic method is time-consuming, it makes sense for clinicians to keep embryos frozen while they await the results. When the results do come in, those embryos with the desired genetic constitution can be thawed and transferred into the woman's uterus.

There is a final use of cryopreservation that is political rather than medical. As evidence of its perceived importance, it was actually the first of five applications listed in a review article by Alan Trounson, an Australian pioneer of reprogenetic technology and the first to facilitate the birth of a baby from a frozen embryo. Trounson writes that cryopreservation provides "a solution to the collection of excess oocytes and the development of more embryos than is required for transfer in human IVF." In other words, to avoid killing embryos, you can freeze them . . . forever. Human embryos take up very little space, and billions—literally—could be stored in a single small tank of liquid nitrogen. With cryopreservation, you can honestly say to the politicians that your clinic does not intentionally bring about the death of embryos during the course of an IVF protocol.

Even before a single child had been born from a frozen embryo, cryopreservation set off a media and political storm with a case of two "orphaned embryos" in the Australian clinic where Alan Trounson worked. A few years later, a "custody battle" in Maryville, Tennessee, demonstrated the sorts of unanticipated legal dilemmas that it could produce. And more recently, an attempt by the British government to regulate the practice with the mandated destruction of 3,000 embryos on the same day shows how confusing the issues can become. It is fascinating to consider the details surrounding each of these stories, and the way they played out in the media, the courts, and ultimately, the IVF clinics where the embryos were stored.

ORPHANED EMBRYOS IN AUSTRALIA

In June 1981, Mario and Elsa Rios traveled from their home in Los Angeles to an IVF clinic at the Queen Victoria Medical Center in Melbourne. They went to Australia because the United States had only a few IVF clinics at that time, and all considered Elsa Rios to be too old for treatment at the age of thirty-seven (which would certainly not be a problem today). Elsa produced three eggs that were fertilized in vitro. One was implanted immediately. Elsa got pregnant, but miscarried soon after. The other two embryos were frozen in liquid nitrogen. Elsa was not ready emotionally to try IVF again, and the couple departed from Australia leaving the two frozen embryos behind.

In April 1983, Elsa and Mario Rios died together in the crash of a small plane near Santiago, Chile. Suddenly, there were "orphaned embryos," and no one knew what to do with them. The Australian IVF clinic had not thought to ask for—and the Rioses had not thought to provide—instructions, in the case of such an eventuality.

At this point, the facts made an interesting news story, but when it was discovered that the Rioses had left behind an estate worth more than $8 million without a will, a media explosion occurred. The headline in the *Daily Telegraph* (in England) read " 'Orphan' Embryo (sic) Heir to Fortune." Women from around the world volunteered themselves as surrogates to bring the embryos to term, dollar signs dancing in their heads.

In response, the government of Victoria (the Australian state in which the IVF clinic was located) stepped in and appointed an independent commission to make a recommendation. In the summer of 1984, the commission reached a decision: the embryos should be thawed and "set aside in the laboratory," a euphemism for their destruction.

The commission reasoned that in the absence of any instructions at all, the Rioses had clearly not given consent to have their children brought to term by another woman. Furthermore, since the embryos had been frozen before techniques of cryopreservation had been perfected, their chances of survival were close to zero anyway.

As might be expected, there was a right-to-life outcry at the thought of these little orphaned embryos being handed a death sentence. And in response to this outcry, the parliament of Victoria passed a law that voided the commission's recommendation and specifically prohibited their destruction.

In the meantime, a California court declared (and the authorities in Victoria agreed) that neither the embryos, nor the children they might become, had the right to claim inheritance of the multimillion dollar estate left behind by the Rioses. Not surprisingly, this greatly dampened interest among potential surrogate mothers. So while the minister of health in Victoria decreed, in 1987, that the embryos should be thawed and transferred to the womb of a volunteer, they still remain in a state of limbo in a tank of liquid nitrogen in the Queen Victoria Medical Center in Melbourne. And there they will probably stay . . . forever.

Some of you may be confused about what ethical dilemma was actually raised in this case. You're not alone. The embryos were produced to help Elsa and Mario Rios have a child whom they could raise as part of a family. When Elsa and Mario died, this reproductive goal could no longer be achieved. The Rioses never intended for someone else to have their child. So it seems logical that the embryos should be thawed and disposed of.

But many politicians are anything but logical or consistent in the way they make policy. Although Australian women can choose to abort—and thus bring about the death of—fetuses that are much more developed than two frozen clumps of cells with little chance of survival under any circumstances, it's the image of those poor little orphans that counts. Unfortunately, such nonsense will continue to prevail as long as there are politicians who fear the wrath of those who believe that embryos are equivalent to human beings.

A CUSTODY BATTLE FOR EMBRYOS
IN TENNESSEE

Mary Sue wanted to be a mother. In 1979, at the age of eighteen, she married Junior Lewis Davis, and over the next four years, she tried repeatedly to achieve a normal pregnancy. She actually got pregnant on five separate occasions, but in each case, the embryo implanted dangerously into a fallopian tube, not her uterus. The last of these ectopic pregnancies ruptured one of her tubes before it could be terminated, and her other tube was severed to avoid further repetitions of the same condition.

At the age of twenty-two, Mary Sue was now infertile. Her only hope for getting pregnant was through the use of IVF. She entered a program directed by Dr. Ray King at the Fertility Center of Eastern Tennessee in Knoxville, and on six separate occasions over a period of four years, she underwent hormonal injections to produce eggs that were retrieved by laparoscopy, fertilized with her husband's sperm, and allowed to develop into embryos before being transferred into her uterus. On all six occasions, the embryos failed to implant.

After an attempt at adoption was scuttled when the birth mother decided to keep the baby, Mary Sue and Junior Lewis entered Dr. King's IVF program one more time. At this point, Dr. King had introduced cryopreservation as a component of his IVF protocol. In December 1988, nine eggs were recovered from the ovaries of the twenty-seven-year-old woman to be fertilized with her husband's sperm. The resulting embryos were allowed to develop in vitro to the four- to eight-cell stage when two were introduced into Mrs. Davis's womb, while the other seven were frozen in liquid nitrogen. Once again, the introduced embryos failed to implant.

Two months later, in February 1989, the Davises separated, and Junior Lewis filed for divorce. The frozen embryos quickly became a matter of contention. Mary Sue wanted to use the embryos in what she considered a last ditch effort to get pregnant. But Junior Lewis had decided that he did not want a child born from the union between him and his estranged wife.

"I consider them life," Mary Sue responded. And to mollify her estranged husband's concerns, she also said that she would raise the child by herself and would not ask for child support.

The divorce trial began on August 7, 1989, in Maryville, Tennessee. On the second day, Junior Lewis testified that he would feel "raped of his reproductive rights" if the embryos were "inserted in Mary or any other donor." He said that he still felt the pain of his own parents' divorce, and he strongly objected to bringing a child into the world under the same conditions. However, he was also strongly opposed to abortion and did not want to see the embryos destroyed. Rather, he asked the court to grant "joint custody" of the embryos to him and Mary Sue, in which case they would remain frozen until both parties could agree on what to do with them. "(It) is a joint decision," Junior Lewis testified. "Her input is just as important as mine. Hopefully she'll learn to understand they are part me as well as part her."

In contrast, Mary Sue asked that the embryos be made available to her immediately so that she could attempt to have them implanted within her uterus. Outside the courtroom, she told reporters, "It's not just his child, it's my child too. They've already been conceived. I feel it's my right to have my child." Her attorney argued that the embryos should be treated as "preborn children" whose best interest would be served by having them develop to term in Mary Sue's body.

Mary Sue desperately wanted to have a child and felt that her only chance in fulfilling this dream lay in the contested embryos that she had produced. Junior Lewis did not want to be a father and felt that his reproductive rights would be violated if Mary Sue implanted the embryos without his permission. How could one choose between these competing views when both appeared to be valid?

One legal argument was that the embryos should be treated as property to be disposed of like other assets in a divorce proceeding. Typically, property can be divided equally between parties or awarded to one party while the other receives monetary compensation. Interestingly, the division of embryos today—in the literal sense—could lead to their multiplication by cloning, which might have satisfied Mary Sue, but not Junior Lewis.

In fact, the trial court judge based his decision not on the desires of Mary Sue, and not on the desires of Junior Lewis, but on the perceived desires of the embryos themselves, as suggested by Mary Sue's attorney. The judge started from the assumption that the embryos were children, not property, and on this basis, he made his ruling from the point of view of a custody dispute. In such a dispute, he said, the best interests of the children are paramount. In this case, the interests of the embryo-children were best served by allowing them to come to term within Mary Sue's body. The fact that this ruling coincided with Mary Sue's desires was of no consequence to the court.

Junior Lewis appealed this decision and it was reversed by a higher court, which appeared to be substantially swayed by testimony presented by the bioethicist John Robertson. Robertson had argued that the case should be decided "in favor of the person who would be hurt worse by losing." By the time the higher court had reached its decision, Mary Sue had remarried and was no longer interested in having the embryos implanted into herself, but now wished to have them donated to another infertile couple. This change in situation clearly tipped the balance in favor of Junior Lewis who had the most to suffer from being forced into parenthood against his will.

It is interesting to ponder what the high court would have done if the original circumstances had remained unchanged and there really was no way to decide which party had the most to lose. Luckily, legal battles like this one and those that surrounded the orphaned Australian embryos are less likely to occur today for a very simple reason. IVF clinics have learned their lesson in the wake of these highly publicized cases, and they now work closely with lawyers to develop contracts that patients must sign before being allowed to proceed with any reprogenetic technology. These contracts clearly spell out how future embryos should be treated in the event of any imaginable scenario or dispute, including death and divorce.

ABANDONED EMBRYOS IN GREAT BRITAIN

So long as decisions are based on agreements reached between patients and their doctors, it would seem that reason could prevail. But sometimes governments try to get into the act as well, and sometimes they cause more problems than they solve. This was the case with the British law passed in 1990 that required "parents" to give their consent to have frozen embryos stay in storage for longer than five years. The law came into effect on August 1, 1991.

On July 31, 1996, the *Washington Post* story began: "Some 3,000 frozen human embryos, essentially abandoned by their parents (sic), face destruction Thursday morning [August 1] under a British law that limits the storage of unclaimed in vitro fertilized eggs." The parental consent requirement was being implemented for the first time on all embryos that had been placed into storage prior to the August 1, 1991, date. In the months before the 1996 deadline, the thirty-three clinics involved had tried to contact all of the couples who had not previously indicated what they wanted to have done with their long-storage embryos. About two-thirds were reached and gave their consent either to have their embryos stored longer, or to have them destroyed quietly before the deadline was reached.

But as of July 31, 650 couples had still not been reached. Since they had obviously not given their consent to further storage, their embryos had to be destroyed according to the law. Unless the law is changed, additional embryos are likely to meet this same fate every week from this point on, as the five-year storage period associated with each draws to a close.

As expected, the Vatican denounced the destruction as "a prenatal massacre," and hundreds of women in Italy, including some nuns, offered to "adopt" the embryos in order to act as their legal guardians. The Catholic Church leadership in Britain argued that if the embryos must be destroyed, they should at least be given "a proper funeral."

Many others, who did not equate embryos with human beings, were equally upset by the British mandate for a different reason. They saw the five-year time limit as entirely arbitrary and contrary to the rights of those who had placed the embryos in storage and could not now be reached for reasons that might be as simple as an address change. This was the very first time that a law had required the destruction of embryos *without* the consent of the couples that had produced them. Peter Brinsden, director of the Bourn Hall IVF clinic set up by Steptoe and Edwards, said that he contemplated going to jail—the penalty for violating the law—rather than destroying embryos without consent.

Some would say that people who really cared about the continued storage of their embryos would have realized what was going on from the massive media blitz and would have gotten in contact with their IVF clinics on their own. By this line of reasoning, most of the unclaimed embryos really were abandoned and not deserving of further storage, while the embryos of concerned "parents" could remain safely frozen. But on August 1, 2001, even concerned "parents" will have no recourse to prevent the death of their embryos. For on that day, the same British law that requires consent for embryo storage longer than five years also mandates the destruction of *all* embryos that have been kept in storage for ten years. This destruction is required by law even if those who produced the embryos request otherwise. In the United States (in 1997), there is no federal law that places any similar restrictions on the practice of IVF, cryopreservation, or any other reprogenetic technology.

IN SUSPENDED ANIMATION

With the ability to keep embryos floating indefinitely in suspended animation, it becomes possible to place large distances in time and space between the producers of the embryos—the genetic parents—and the children that emerge from them.

This raises the fanciful possibility of parents arranging for their children to be born long after they are gone. Most people would not want to do this. Most choose to become parents so that they themselves can experience the lives of their children. But a few may revel in the idea of a kind of time travel. While they can't go into the future, they can put their children there. They could establish a trust with the legal authority to look after their frozen embryos and to pay handsome sums of money to a future woman who was willing to act as a surrogate to bring them to life, and to others willing to act as foster parents. Perhaps they would arrange for their children to be born near the end of the third millennium, in the year 2998.

Or we can go back two thousand years and imagine what could have happened if this technology had been available to Cleopatra and her lover, Julius Caesar. Their children could be living among us today.

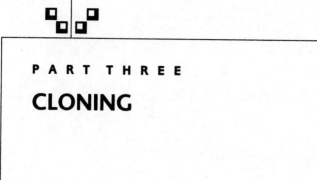

PART THREE

CLONING

Let us make man in our image, after our
likeness.

—GENESIS 1:26

8

From Science Fiction
to Reality

FEBRUARY 23, 1997

On the last Sunday in the month of February, in the third year before the end of the second millennium, the world woke up to a technological advance that shook the foundations of biology and philosophy. On that day, we were introduced to Dolly, a six-month-old lamb who had been cloned directly from a single cell taken from the breast tissue of an adult donor.

There were lead stories on every television and radio news broadcast and headline banners on the front page of every newspaper around the world. And for weeks afterward, it didn't let up. Story after story came out discussing the stunning implications of this monumental achievement. On the streets, in offices, on campuses, and in classrooms, people couldn't stop talking about it. One little lamb had succeeded in changing our conception of life forevermore.

Perhaps more astonished than any of their neighbors were the scientists who actually worked in the field of mammalian genetics and embryology. Outside the lab where the cloning had actually taken place, most of us thought it could never happen. Oh we would say that perhaps at some point in the distant future, cloning might become feasible through the use of sophisticated biotechnologies far beyond those available to us now. But what we really believed, deep in our hearts, was that this was one biological feat we could never master. New life—in the special sense of a conscious being—must have its origins in an embryo formed through the merger of gametes from a mother and father. It was impossible, we thought, for a cell from an adult mammal to become reprogrammed, to start all over again, to generate another entire animal or person in the image of the one born earlier.

How wrong we were.

Of course, it wasn't the cloning of a sheep that stirred the imagination of billions of people. It was the idea that humans could now be cloned as well in a manner akin to taking cuttings from a plant, and many people were terrified by the prospect. Ninety percent of Americans polled within the first week after the story broke felt that human cloning should be banned. And the opinions of many media pundits, ethicists, and policymakers, though not unanimous, seemed to follow those of the general public. The idea that humans might be cloned was called "morally despicable," "repugnant," "totally inappropriate," as well as "ethically wrong, socially misguided, and biologically mistaken."

Many of the scientists who work directly in the field of animal genetics and embryology were dismayed by all the attention now directed at their research. Most unhappy of all were those associated with the biotechnology industry, which has the most to gain in the short-term from animal applications of the cloning technology. Their fears were not unfounded. In the aftermath of Dolly, polls found that two out of three Americans considered the cloning of *animals* to be morally unacceptable, while 56 percent said they would not eat meat from cloned animals. The British government decided to "reward" the scientist actually responsible for Dolly's creation, Ian Wilmut, with the *withdrawal* of all further funds for his research. Clearly, nervous politicians wanted to distance themselves as far as possible from his controversial achievement.

It should not be surprising, then, that many scientists in the field tried to play down the possibility of human cloning. First they said that it might not be possible *at all* to transfer the technology to human cells. And even if human cloning is possible in theory, they said, "it would take years of trial and error before it could be applied successfully," so that "cloning in humans is unlikely any time soon." And even if it becomes possible to apply the technology successfully, they said, "there is no clinical reason why you would do this." And even if a person wanted to clone himself or herself or someone else, he or she wouldn't be able to find trained medical professionals who would be willing to do it.

That's not what science, history, or human nature suggest to me. The cloning of Dolly broke the technological barrier. There is no reason to expect that the technology couldn't be transferred to human cells. On the contrary, there is every reason to expect that it *can* be transferred. It requires only equipment and facilities that are already standard or easy to obtain by biomedical laboratories and free-standing in vitro fertilization clinics across the country and across the world. Although the protocol itself demands the services of highly trained and skilled personnel, there are thousands of people with such skills in the United States alone.

It is not a question of whether human cloning will work, but whether it could be used safely or not. Historical precedent suggests that reprogenetic service providers may not even wait until this question has been resolved. The direct injection of sperm into eggs (ICSI) as a cure for infertility was embraced by the IVF community as soon as the technique was perfected, long before any consequences to the children born could be ascertained. And as we shall see, the demand for cloning from individuals and couples is sure to be stronger than the demand for ICSI.

Before we take a closer look at who might want to use human cloning as means for reproduction, and what their reasons are, it is worthwhile to start at the beginning with answers to some basic questions. What is a clone? How was Dolly made? And why does she terrify so many people, even as she thrills a few others?

FROM PLANTS TO TADPOLES, BUT NOT MICE

The word *clone* first appeared in the language of science at the beginning of the twentieth century to describe "groups of plants that are propagated by the use of any form of vegetative parts." Since that time, *cloning* has been used to describe the process by which a cell, or group of cells, from one individual organism is used to derive an entirely new organism, which, according to the definition, is a "clone" of the original. When multiple individuals are cloned from a single ancestor, they are all considered to be "members of a clone." The critical defining characteristic of a cloned individual is that it is *genetically identical* to the ancestral cell or organism from which it is derived, as well as to any other clones derived from the same ancestor.

Among single-cell organisms like bacteria, cloning is as natural as can be. When bacterial reproduction takes place through cell division, the two daughter cells are clones of each other. Plants, on the other hand, normally reproduce sexually through the production of fertilized seeds that contain new combinations of genetic material not found in their parents. With human intervention, however, most plants can easily be cloned through the use of *cuttings* or bulbs—vegetative parts—from "donor parents."

The word *clone* would never have entered the public lexicon if it had remained in the provenance of plants and microbes. However, in the 1960s, the attempts of a British embryologist named John Gurdon to clone a vertebrate animal—the frog—reached the eyes and ears of the media.

The cloning of animals had to proceed in a very different manner from the cloning of plants. It is not possible simply to take a cell from an adult, place it in an embryonic environment, and then expect it to revert to an embryonic form from which a whole new animal could develop. The reason this approach won't work is that animal cells are much less flexible than plant cells in terms of their developmental potential. Plants always develop in response to their environment, and even when two plants have identical genetic material, they grow into very different structures. In addition, many differentiated plant cells have the capacity to transform themselves into totally different types. So when a branch is cut off from one plant and provided with water, it can sprout new roots and become a whole new plant.

In contrast, differentiated animal cells are greatly restricted in their developmental capacity, as we discussed in chapter 4. Each cell in the body of an adult is committed to a particular function. No normal adult cell—other than a sperm or an egg—has the ability to transform itself into a completely different type of cell. Liver cells cannot become brain cells and skin cells cannot be transformed into early embryo cells. Why is this so, you may ask, when every cell has the same genetic material? If the genetic material is all there, it might seem as if there should be some way to convert an adult cell back into an embryonic cell.

The problem is that each type of cell looks the way it does and performs the functions that it does because it is programmed to "read" only a well-defined portion of its total genetic material. The programming is accomplished by the presence of hundreds or thousands of special protein signals that sit securely on the DNA, instructing some genes to function and other genes to remain silent. In order for a skin cell to become converted into an embryonic cell, its entire genetic program would have to be altered in a particular way, and this could only be accomplished by a massive, but highly precise, substitution and reshuffling of the protein signals that are attached to the genetic material.

In theory, the simplest way to get around this problem would be to extract the genetic material from a single skin cell, strip away its associated protein signals, and then place this genetic material inside an egg cytoplasm whose own genetic material had been previously removed. The egg cytoplasm contains all of the particular protein signals required for starting the embryonic program of gene expression. These signals would hop onto the naked DNA and development would be initiated into a clone of the individual who donated the adult genes.

There are major technical problems that make this approach daunting, if not impossible. One is that the skin cell's signals are tightly bound to the genetic material and not easily removed. A second, more serious problem is that naked genetic material of the size present in animal cells always breaks apart when it is handled, no matter how gently. And when DNA molecules break, they cannot be transmitted accurately to daughter cells with each cell division. Thus, even if it were possible to place all the DNA from a single skin cell into an egg cytoplasm, the resulting embryo would have no chance of developing into an adult animal.

So, at the very start, scientists decided to use the next best thing to naked DNA—an isolated single-cell nucleus with a membrane that acts as a protective shield against chromosome damage when it is picked up from one cell and placed into another. The unavoidable downside of this approach is that some of the original cell's signals—those attached to the DNA and others present in the nucleus—are brought along with the foreign genetic material into the cytoplasm of the embryonic cell.

The use of "nuclear transplantation" as a means toward the cloning of animals was first developed by Robert Briggs and Thomas King working at the Institute for Cancer Research in Philadelphia during the early 1950s. The frog was chosen for these experiments because its eggs are very large and readily accessible to manipulation. Although Briggs and King never reached their goal of cloning from adult cells, they set the stage for John Gurdon, who finally succeeded in using this method to obtain tadpoles during the mid-1960s.

The cloning of frogs was never easily accomplished. After transplanting thousands of nuclei extracted from adult skin and gut cells, Gurdon's success rate was still abysmally low, and the few animals he obtained developed only to the tadpole stage before dying. It is certainly possible— and with hindsight, it now seems likely—that Gurdon's difficulties were mainly a consequence of the primitive equipment and technology available at that time. For even a small amount of damage to nuclei or reconstructed eggs could have drastic consequences on development.

But, most scientists interpreted Gurdon's essentially negative results differently. Rather than blaming the technology, we blamed mother nature herself. In an almost religious way, we assumed the existence of a basic biological principle: adult cell nuclei cannot be readily reprogrammed back to an embryonic state. The rare adult donor nucleus that did turn into a tadpole was presumed to have come from an aberrant cell. And if a tadpole could be obtained only rarely, it seemed reasonable to assume that it would never be possible to clone adult cells of more highly developed mammalian species—like human beings—into healthy live-born children.

Indeed, in 1984, when the highly respected embryologist Davor Solter, and his student James McGrath, reported on an extensive series of nuclear transplantation studies—with better equipment and technology—on mouse eggs, their results seemed to validate this basic biological principle. The concluding sentence of their publication in the journal *Science* stated that "the cloning of mammals by simple nuclear transfer is biologically impossible."

CLONING ENTERS PUBLIC CULTURE

Although scientists viewed Gurdon's results in one light, popularizers of science viewed it in quite another. The fact that even a single frog had been cloned led to the suggestion that cloning *would* be possible with human beings. The idea began to filter into the public consciousness during the late 1960s and was firmly planted there with the 1970 publication of Alvin Toffler's sensational, and still influential, *Future Shock*. Toffler wrote, "One of the more fantastic possibilities is that man will be able to make biological carbon copies of himself. . . . Cloning would make it possible for people to see themselves anew, to fill the world with twins of themselves. . . . There is a certain charm to the idea of Albert Einstein bequeathing copies of himself to posterity. But what of Adolf Hitler?"

Just as the concept of cloning was being absorbed by the public, it was parodied by Woody Allen in his 1973 movie *Sleeper*. Allen plays the mild-mannered Miles Monroe who is transported two hundred years into the future and is mistaken for the chief surgeon charged with the task of bringing back the recently deceased "Leader" of the country. While the Leader has met with an untimely death, his nose has been kept alive for nearly a year through a "massive biochemical effort." Miles Monroe is supposed to clone The Leader's whole body from his nose, as the top biomedical scientists of this future country watch from an operating room observation deck. Allen toys with the dual meaning of life—cellular versus conscious—when his character kidnaps the nose and threatens to shoot it if he is not allowed to go free.

Five years later, the 1978 movie *The Boys from Brazil*, based on a book by Ira Levin, took up Toffler's more menacing idea of a Nazi plot to clone an army of latter-day Adolf Hitlers. And that same year, the J. B. Lippincott Company published a supposed nonfiction book by the science writer David Rorvik entitled *In His Image: The Cloning of a Man*. Rorvik claimed to tell the story of a "worldly, self-educated, aging millionaire" who wanted an heir and succeeded in obtaining "not exactly a son," but rather his genetic equivalent through the use of the same nuclear transplantation technique that John Gurdon had used to clone frogs. Rorvik never provided evidence in support of his claim, and several years later his publisher was forced to admit the book was a hoax.

By the early 1980s, the notion of cloning had become entrenched in popular culture, appearing again and again in movies, television shows, and science fiction novels. And it entered the inanimate world as well, with clones of computers and even perfumes. Clones were seen as almost, but not quite, perfect copies of the original, usually cheaper and assumed to be not as "sharp" in some way.

But even as clones flooded the popular imagination, very little in the way of new scientific results was publicized. Some people knew that frogs had been cloned, but it seemed that no real scientific progress had been made beyond that organism. And then in 1993, the silence was shattered with a report that two George Washington University scientists, Jerry Hall and Robert Stillman, had "cloned human embryos."

The Hall-Stillman experiment caused a brief media stir far out of proportion to what had actually been accomplished. Hall and Stillman had simply taken seventeen early human embryos, between the two-cell and eight-cell stages, removed their zona coats, and then separated each of the cells in each embryo apart from its neighbors. Each individual cell was next surrounded by a synthetic zona coat and allowed to develop in a laboratory dish by itself. After a few days, Hall and Stillman found forty-eight newly formed embryos developing in a normal manner. The experiment was terminated at this point—out of ethical consideration—and the embryos were discarded.

Embryo cloning is a far cry from adult cell cloning. If the Hall-Stillman experiment had been taken to its logical endpoint, it might have been possible to obtain the birth of identical twins or triplets. But even the normal practice of IVF results in the birth of twins or triplets, albeit nonidentical ones. And the old-fashioned method of reproduction through intercourse produces a million pairs of newborn identical twins, a lesser number of identical triplets, and perhaps a handful of identical quadruplets, around the world each year. So what Hall and Stillman had accomplished in the laboratory was equivalent to a well-known natural process.

Still, even this mimicry of nature provoked immediate outrage from many political corners. The Vatican called it a "perverse choice" and a "venture into a tunnel of madness." Biotech critic Jeremy Rifkin said it heralded "the dawn of the eugenics era," and he organized protest rallies outside the institution where it had taken place. The European Parliament voted unanimously to ban cloning because it was "unethical, morally repugnant, contrary to respect for the person, and a grave violation of fundamental human rights which cannot under any circumstances be justified or accepted." And this was all because two scientists had gently teased single embryos apart into two, three, or four separate cells that grew for a few days by themselves before fading away.

I suspect that if the word *clone* had not been used to describe what Hall and Stillman had done, the media would never have jumped on the story. As it was, two weeks passed between their presentation at a scientific meeting and the first headline: "Scientist Clones Human Embryos and

Creates an Ethical Challenge." It was the ominous juxtaposition of those two words—*clones human*—that brought on the hysteria.

FROM EMBRYOS TO ADULTS

While the cloning of Dolly from an adult cell was unquestionably a giant leap forward in reproductive technology, it was a leap that began from a sturdy platform of technical advances that built quietly upon one another over the preceding fourteen years. The first step was accomplished at the Wistar Institute in Philadelphia in 1983, where Davor Solter and Jim McGrath established a protocol for transferring nuclei from one mouse embryo to another. Their work was critically important for two reasons. First, it demonstrated the general feasiblity of using nuclear transfer technology in mammals. Second, it introduced a modification of the technique used in frogs that greatly increased the rate of embryo survival. Instead of isolating nuclei away from their cellular encasement, as Gurdon had done, Solter and McGrath chose to keep nuclei properly protected within their cytoplasmic environments surrounded by a cellular membrane.

The actual protocol began with the removal and elimination of the nuclei that were already present within the recipient embryo. Then the donor cell was placed in the space between the zona coat and the embryo itself, and the two cells were induced to fuse with a special chemical agent or an electrical pulse.

Although they referred to this protocol as "nuclear transplantation"— and it has been referred to in this manner ever since—Solter and McGrath never transplanted nuclei directly into recipient embryos. Rather, they implanted donor cells next to embryos and then allowed a fusion event to bring the donor nucleus into the cytoplasm of the recipient cell. By keeping donor cells intact until the moment of fusion, Solter and McGrath succeeded in protecting the genetic material within. Their protocol was so efficient and safe that 90 percent of embryos reconstructed with nuclei from other early embryos survived and developed properly.

The next advance on the way to Dolly was accomplished in 1986 by Steed Willadsen, who was working at the ARFC Institute of Animal Physiology in Cambridge, England. What Willadsen did differently from Solter and McGrath was to use nuclear-free *unfertilized* eggs, rather than one-cell embryos, as recipients for donor nuclei. The logic behind this decision is based on the notion that an unfertilized egg is chock full of signal proteins waiting patiently to pounce onto the naked DNA that it expects to receive from the fertilizing sperm cell. And if the egg is presented with a donor nucleus instead, the egg's signal proteins won't know the difference— they'll blindly *try* to pounce onto the donor cell DNA with the same vengeance. This logic was validated when Willadsen reported the birth of healthy lambs that had been cloned from donor cells derived from 8-cell embryos.

Eight more years went by before another important advance in cloning was made by Neal First at the University of Wisconsin in 1994. This time the species was the cow, the donor cells were obtained from an even later embryonic stage, and four calves were born. What First didn't realize, however, was the probable reason for his success. It turns out that a technician in First's laboratory had mistakenly not provided the donor embryo cells with nourishing serum that all cells need to grow properly. As a result, the donor cells stepped out of their normal cycle of growth and division and paused in a type of hibernation phase known to scientists as G0. Could it be that cells in this special state of hibernation might be more amenable to cloning than other cells? Perhaps the signal proteins sitting on the DNA in these cells are more easily dislodged by the ones waiting in the egg cytoplasm.

Keith Campbell and Ian Wilmut at the Roslin Institute in Edinburgh, Scotland, were intrigued by this possibility, and they set about trying to test it with their favorite animal, the sheep. They easily obtained lambs after nuclear transplantation from nine-day-old embryo donor cells, and they extended their success to donor cells obtained from embryo-like cultures grown over a period of weeks in a laboratory dish. They reported their results in a March 1996 paper entitled "Sheep Cloned by Nuclear Transfer from a Cultured Cell Line." And then they moved on to more advanced donor cells, using precisely the same techniques.

Dolly was born at 5:00 P.M. in the afternoon on July 5, 1996. She resulted from the fusion of a nuclear-free unfertilized egg with a donor cell obtained from the mammary gland of a six-year-old ewe. She was the first mammal to be cloned from an adult cell, and is a generation removed from the fertilization event that actually brought together the gametes from her genetic parents.

Dolly's existence was announced to the scientific community in a paper published in the journal *Nature* on February 27, 1997. Unnoticed in the commotion surrounding this one lamb is the fact that two others were also cloned from skinlike cells obtained from a fetus. The birth and survival of three healthy lambs from highly differentiated donor cells provides a clear demonstration that the cloning of a lamb was not a fluke.

9

Human "Cuttings"

FROM SHEEP TO PEOPLE?

Dolly is only a ewe, and as I write these words, a human being has yet
to be cloned. What are the chances that it will happen? How likely is it
that the technology developed in sheep could be transferred to our own
species? And how quickly could it come about?

Answers to these questions are grounded in the understanding that
embryos of all mammals undergo early development in a very similar way,
although there are small differences among species. If cloning is more
dependent on the similarities among mammals than their differences, then
human cloning will be possible. This proposition can be tested through
attempts to clone a variety of mammalian species.

As far as the technique of nuclear transfer goes, the results are already
in. Even before the announcement of Dolly, scientists at many institutions
had succeeded in producing cows, pigs, goats, rabbits, and mice from
embryos with transplanted nuclei. And within a week after the Dolly an-

nouncement, scientists at the Oregon Primate Research Center in Beaverton reported the first successful nuclear transfer in a primate species, the rhesus monkey. If nuclear transplantation works in every mammalian species in which it has been seriously tried, then nuclear transplantation *will* work with human cells. The monkey result, in particular, is the clincher because humans are nothing more than glorified monkeys when it comes to embryonic development.

At the time of writing, however, Dolly remains the only animal born after nuclear transfer from an *adult* donor cell. There is no reason to expect that adult *human* cells won't make good nuclear donors, but we won't know for sure until experiments with other species (especially monkeys) are completed, which will almost certainly come to pass within the next several years.

For human beings, though, it's not just a question of whether it *could* work, it's a question of whether it could work *safely*. A basic principle of medical ethics is that doctors should not perform any procedure on human subjects if the risk of harm is greater than the benefit that might be achieved. In the case of cloning, this principle would oblige physicians to refrain from practicing the technology unless they were sure that the risk of birth defects was no greater than that associated with naturally conceived children.

Many of the media reports that described the birth of Dolly emphasized the fact that the success rate was only 1 in 277 "tries." The implication—sometimes stated explicitly—was that many lambs died or were born with genetic malformations. But, this is a misunderstanding of what the number 277 actually represented in the published report. What it stood for was the number of fusions that were initially obtained between donor cells and unfertilized eggs. Only 29 of these fused cells actually became embryos, and these 29 embryos were introduced into 13 ewes, of which one became pregnant and gave birth to Dolly. If safety is judged by the proportion of those lambs born who were in good health, then the record to date is perfect (albeit with a rather small sample size).

In fact, there is no scientific basis for the belief that cloned children will be any more prone to genetic problems than naturally conceived children. The most common type of genetic birth defect results from the presence of an abnormal number of chromosomes. Trisomy 21, responsible for Down syndrome, is the most prevalent example. Abnormalities of this class are caused by mistakes that occur when the genetic material is reduced by half during the process of sperm or egg formation. With cloning, there is no reduction in genetic material, and the chance that mistakes of this type will occur is greatly reduced.

The second most common class of genetic abnormalities results from the inheritance of two mutant copies of a gene that were each carried silently within the two parents. Examples from this class include Tay-Sachs disease, sickle cell anemia, cystic fibrosis, and PKU. With cloning, any silent mutation in the donor will remain silent within the newly formed embryo and child as well.

Finally, much less frequently, a new mutation can occur in the genetic material of the egg or sperm and lead to a birth defect in the person that is born. With cloning, there is the same low probability of a new mutation in the genetic material brought in with the donor nucleus.

Surprisingly, what all of these comparisons suggest is that birth defects in cloned children could occur less frequently than birth defects in naturally conceived children. There is, however, no way to predict whether cloned children would be just as healthy, as a group, as all other children until direct data on large numbers of cloned animals are obtained. Experiments are now underway to clone many other species from adult cells. And an answer to the question of cloning's effect on health and aging is likely to come most quickly with small animals like mice that have a naturally short life span. If cloning is found to have no effect on the health or life span of experimental animals, it would be reasonable to conclude that the same would hold true for human beings. And with this conclusion, a major—if not *the* major—objection to human cloning will be eliminated.

Even if experiments in animals demonstrate the safety of cloning, there is still a question of feasibility. Critics point out, once again, that only 1 lamb was born after 277 attempts and conclude that the efficiency of the protocol is so low as to make it impractical for use with human beings. Once again, though, the critics are mistaken. No matter how you look at the numbers, they are better than those obtained during the initial development of human IVF. Steptoe and Edwards worked with hundreds of eggs, over more than a decade, to perfect the process of fertilization in a laboratory dish. They then introduced embryos into women on dozens of occasions before achieving the birth of Louise Brown. And for years after this first birth, the average IVF success rate—calculated as the proportion of women who gave birth after receiving embryos—was *less* than the 1 in 13 reported in the Dolly experiment. Nevertheless, thousands of couples were willing to spend tens of thousands of dollars on the off chance that they would be the ones who went home with a baby.

With time, of course, technical improvements have increased the efficiency of IVF, and there is every reason to believe that the efficiency of cloning could be similarly improved as long as experimentation along these lines is allowed. Indeed, within six months of the announcement of Dolly's birth, in August 1997, a small Wisconsin company reported the birth of a group of entirely healthy calves cloned by a new improved protocol one hundred times more efficient than the Wilmut approach. The implications for the feasibility of human cloning are as clear as can be.

Once questions of safety and efficiency are eliminated, will there be medical professionals who would be willing to do it? Of this, there can be no doubt. Many IVF clinics already perform sperm injection into unfertilized eggs (ICSI), which uses the same equipment and differs only in detail from the cloning protocols described to date. With a little practice on some spare eggs, skilled IVF practitioners could quickly master the published techniques and improve upon them. And within three weeks of the announcement of Dolly's birth, I learned through casual conversation of two prominent IVF practitioners in different countries who were anxious to move ahead with selected "patients." If, without even looking, I could find two doctors who were willing to clone people, imagine how many more must be out there among the thousands who have the skills to do it.

CLONING MISPERCEPTIONS

Why do four out of five Americans think that human cloning is "against God's will" or "morally wrong"? Why are people so frightened by this technology? One important reason is that many people have a muddled sense of what cloning is. They confuse the popular meaning of the word *clone*, and the specific meaning it takes on in the context of biology.

In its popular usage, *clone* refers to something that is a duplicate, or cheaper imitation, of a brand-name person, place, or thing. The British politician Tony Blair has been called a clone of Bill Clinton, and an IBM PC clone is not only built like an IBM PC, it *behaves* like an IBM PC. It is this popular meaning of the word that caused many people to believe that human cloning would copy not just a person's body, but a person's consciousness as well. This concept of cloning was at the center of the movie *Multiplicity*, which was released just months before the Dolly announcement. In it, a geneticist makes a clone of the star character played by Michael Keaton and explains that the clone will have "all of his feelings, all of his quirks, all of his memories, right up to the moment of cloning." The clone himself says to the original character, "You are me, I am you." It is this image that Jeremy Rifkin probably had in mind when he criticized the possible application of the sheep cloning technology to humans by saying, "It's a horrendous crime to make a Xerox (copy) of someone."

But this popular image bears absolutely no resemblance to actual cloning technology, in either process or outcome. Scientists cannot make full-grown adult copies of any animal, let alone humans. All they can do is start the process of development over again, using genetic material obtained from an adult. Real biological cloning can only take place at the level of the cell—life *in the general sense*. It is only long after the cloning event is completed that a unique—and independent—life *in the special sense* could emerge in the developing fetus. Once again, it is the inability of many people to appreciate the difference between the two meanings of "life" that is the cause of confusion.

A second reason people fear cloning is based on the notion that a clone is an imperfect imitation of the real thing. This causes some people to think that—far from having the same soul as someone else—a clone would have no soul at all. Among the earliest popular movies to explore this idea was *Blade Runner*, in which synthetic people were produced that were just like humans in all respects but one—they had no empathy. (Coincidentally, *Blade Runner* was based on a 1968 book by Philip K. Dick entitled *Do Androids Dream of Electric Sheep*.) And the same general idea of imperfection is explored in *Multiplicity* when a clone of the Michael Keaton character has himself cloned. The clone of the clone is a dimwitted clown because, as the original clone says, "Sometimes you make a copy of a copy and it's not as sharp as the original."

The Irvine, California, rabbi Bernard King was seriously frightened by this idea when he asked, "Can the cloning create a soul? Can scientists create the soul that would make a being ethical, moral, caring, loving, all the things we attribute humanity to?" The Catholic priest Father Saunders suggested that "cloning would only produce humanoids or androids—soulless replicas of human beings that could be used as slaves." And Brent Staples, a member of the *New York Times* editorial board, warned that "synthetic humans would be easy prey for humanity's worst instincts."

Yet there is nothing synthetic about the cells used in cloning. They are alive before the cloning process, and they are alive after fusion has taken place. The newly created embryo can only develop inside the womb of a woman in the same way that all embryos and fetuses develop. Cloned children will be full-fledged human beings, indistinguishable in biological terms from all other members of the species. Thus, the notion of a soulless clone has no basis in reality.

When the misperceptions are tossed aside, it becomes clear what a cloned child will be. She, or he, will simply be a later-born identical twin—nothing more and nothing less. And while she may go through life looking similar to the way her progenitor-parent looked at a past point in time, she will be a unique human being, with a completely unique consciousness and a unique set of memories that she will build from scratch.

To many people, the mere word *clone* seems ominous, conjuring up images from movies like *The Boys From Brazil* with evil Nazis intent on ruling the world. How likely is it that governments or organized groups will use cloning as a tool to build future societies with citizens bred to fulfill a particular need?

THE *BRAVE NEW WORLD* SCENARIO

> *"Bokanovsky's Process," repeated the Director. . . . One egg, one embryo, one adult—normality. But a bokanovskified egg will bud, will proliferate, will divide. From eight to ninety-six buds, and every bud will grow into a perfectly formed embryo, and every embryo into a full-sized adult. Making ninety-six human beings grow where only one grew before. Progress. . . . Identical twins—but not in piddling twos and threes as in the old viviparous days, when an egg would sometimes accidentally divide; actually by dozens, by scores at a time. . . . "But, alas," the Director shook his head, "we can't bokanovskify indefinitely." Ninety-six seemed to be the limit; seventy-two a good average.*

Thus did Aldous Huxley present one of the technological underpinnings of his brave new world where cloning would be used "as one of the major instruments of social stability." With cloning, it was possible to obtain "standard men and women; in uniform batches. The whole of a factory staffed with the products of a single bokanovskified egg."

Brave New World evoked powerful feelings within people not only because they could see inklings of the rigid conformity of the brave new world society within their own, but because the science was presented in a hyperrealistic manner. Even the most minor technical details were carefully described.

Huxley, for one, was convinced that political forces would evolve in the direction he described. In the foreword to the 1946 edition, he wrote: "It is probable that all the world's governments will be more or less completely totalitarian even before the harnessing of atomic energy; that they will be totalitarian during and after the harnessing seems almost certain." It was the *science* that he was less certain of.

Yet, like so many other twentieth-century intellectuals, Huxley underestimated the power of technology to turn yesterday's fantasy into today's reality. Only sixty-four years after he speculated on the possibility of human cloning, it is on the verge of happening. But now that one aspect of science has caught up to *Brave New World*, what can we say about the politics? Will there be governments that choose to clone?

Definitely not in a democratic society for a very simple reason. Cloned children cannot appear out of the air. Each one will have to develop within the womb of a woman (for the time being). And in a free society, the state cannot control women's bodies and minds in a way that would be necessary to build an army of clones.

But what about a totalitarian government that wanted to produce clones to serve its own social needs: "Standard men and women; in uniform batches. The whole of a factory staffed with the products of a single bokanovskified egg."

This scenario is highly improbable. First, only an extremely controlling totalitarian state would have the ability to enslave women *en masse* to act as surrogate mothers for babies that would be forcibly removed and raised by the state. Ruling governments this extreme are rare at the end of the twentieth century. But even if one did emerge, it is hard to imagine why it would want to clone people.

Would it be to produce an army of powerful soldiers? Any government that could clone would certainly get more fighting power out of high-tech weapons of destruction than even the most muscular and obedient soldier clones.

Would it be to produce docile factory workers? Cloning is not necessary for this objective, which has already been reached throughout many societies. And mind control could be achieved much more effectively with New Age drugs targeted at particular behaviors and emotions (another prediction made by Huxley).

Would it be to produce people with great minds? It is not clear how a government would choose a progenitor for such clones, or what it would do during the twenty years or so that it took for clones to mature into adults. After all that time, a new set of leaders might decide that the wrong characteristics had been chosen for cloning. A better approach would be to simply build a superior system of public eduction that allowed the brightest children to rise to the top, no matter where on society's ladder they began their lives.

In the end, one is hard-pressed to come up with a single strategic advantage that any government might get from breeding clones rather than allowing a population to regenerate itself naturally. Thus, the Huxleyan use of cloning as a means for building a stable society seems very unlikely. But there is an obvious exception—one that could occur in a state or society controlled by a single egomaniacal dictator with substantial financial and scientific resources.

The example that comes to mind is that of the Japanese cult leader Shoko Asahara. Asahara's group, Aum Shinrikyo, included well-educated chemists who produced nerve gas for the purpose of holding the Japanese government hostage. The group was exposed, and their leader was arrested and put on trial after a lethal gas attack on the Tokyo subway system in March 1995. Based on what we have learned about the group, it is possible that it might have had both the financial and technical resources required to put together the facility and equipment needed for cloning, as well as the power of persuasion required to convince skilled personnel to carry it out. And the aura that Asahara projected was such that he might well have succeeded in convincing women to become pregnant with his clones. Finally, Asahara himself seems to have been exactly the kind of egomaniac who would have preferred child clones over naturally conceived sons.

I doubt that we could stop people like Shoko Asahara from cloning themselves. But would it make any difference? Let us imagine that Asahara had cloned himself into a dozen children. It seems extremely unlikely that these children would have any greater effect on society, twenty years down the road, than sons conceived the old-fashioned way. It's not only that they wouldn't grow up in the same adverse environment that played an important role in turning Asahara into the cult leader that he became. It's also that they would grow up among different people who would be unlikely to respond to them in exactly the same way that people responded to Asahara. The same could be said for modern-day clones of Adolf Hitler. In both cases, the original men were catapulted into positions of leadership through chance personal or historical events that will never repeat themselves. An adult alive today with Adolf Hitler's mind, personality, and behavior would be more likely to find himself barricaded in a militia outpost or in jail than in the White House or the German Bundesrat.

While Hitler's Third Reich and Asahara's Aum Shinrikyo were both short-lived phenomena, there are still examples of royal families—albeit with little real power today—that have handed down the crown from parent to child over hundreds of years. If after ascending to the throne, Prince Charles of Great Britain decided to place his clone—rather than his eldest son—next in line, would that upset the world order? On the contrary, I doubt if anyone would care.

Egomaniacs will not be the only people who will want to employ cloning for their personal use. In addition, there will be many more individuals and couples who will quietly choose cloning to satisfy their particular reproductive goals. What those goals might be are considered in the following sections of this chapter.

THE CLONING OF CHILDREN

Anissa Ayala was a sophomore at Walnut High School in a suburb of Los Angeles, California in the spring of 1988 when she was diagnosed with myelogenous leukemia, a slowly progressing but ultimately fatal cancer of blood stem cells. The only way to cure this cancer is through a two step process, the first step entailing a treatment with highly toxic chemicals that destroy all of the blood stem cells—including the cancerous ones—throughout a person's body. Unfortunately, blood stem cells are needed to replenish the differentiated blood cell pool each day as older cells die off. In the absence of stem cells, a person would slowly run out of the differentiated cells that she needs to transport oxygen from the lungs to all her organs, and although she would be cancer-free, within a few days, she would die.

This is why the second step of the treatment process is crucial: replacement of the eliminated blood stem cells with new ones provided by a donor. Since blood stem cells are located in the bone marrow, this second step is accomplished with a bone marrow transplant. But the transplant can only come from a donor who shows good tissue compatibility with the person in need, and the chance of good compatibility between two unrelated individuals is only 1 in 20,000.

The parents of Anissa Ayala—Mary and Abe—were desperate to save her life, and with the help of their family and community, they searched for a donor. None of the members of their extended family were compatible, and a nationwide search conducted over a two year period was also negative. Time was running out.

At this point, the Ayalas made a decision. Mary would try to have another child who could provide Anissa with the needed bone marrow. The odds were heavily against the Ayalas. Abe was forty-five years old and had been vasectomized long before. Mary was forty-two, an age at which many women become grandmothers, not mothers. Furthermore, even if they were able to conceive and bring a child to term, the chances that the child would be a good match with Anissa were only 25 percent. And even if the child were a good match, this would still give Anissa only a 70 percent chance of surviving after the transplant.

Abe and Mary were willing to fight the odds. Abe was able to get his vasectomy reversed, and Mary was able to become pregnant and stay that way for nine months. When amniocentesis was performed, the fetus was found, against the odds, to be compatible with Anissa. On April 2, 1990, Mary Ayala gave birth to Marissa Eve. Fourteen months later, the bone marrow transplant from Marissa to Anissa was performed.

On June 9, 1996, Anissa—healthy and cancer-free—and her sister Marissa sat together with their parents for a CNN television interview to celebrate the five-year milestone at which a cancer patient is considered to have beaten her illness. The six-year old Marissa positively beamed when she told the interviewer, "I saved her (Anissa's) life." After the show, Abe and Mary gave big hugs to both of their very-much-loved daughters.

Let's now consider what Mary and Abe might have done if cloning had been available to them when they first learned of Anissa's illness. By starting with a skin cell from Anissa's body, they would have been able to construct a new embryo with the same genetic material. And instead of fighting the odds, they would have known from the outset that their new child would be not just a good donor, but a *perfect* match for their older daughter (which is *only* possible with identical twins). Except in one detail, the end result would be no different from the one that actually happened. A child named Marissa would still have been born, and she would still have cured her older sister. The one difference would be, of course, that instead of having genetic material that was 99.95 percent the same as her older sister like other nontwin sibling pairs, Marissa's genetic material would be 100 percent the same as Anissa's. Would that make any difference in the amount of love that Marissa's parents gave to her? Would she be any less proud of saving her sister's life?

It is interesting to look at how some prominent bioethicists reacted when the Ayala story first broke in the news media six years earlier, when Marissa was still a fetus in her mother's body. The general reaction was one of outrage. "It's absolutely ethically wrong to have a child as a donor, just for what it (the child) can do for someone else," said Arthur Caplan. "You are treating a human being as an object and that takes away from the value of the infant itself," said Reinhard Preister. "Children are not medicine for other people," said George Annas. "What they're doing is ethically very troubling," said Alexander Capron. When asked to respond directly to a scenario in which an older child could only be saved with the use of cloning, Richard McCormick, a Jesuit priest and professor of Christian ethics at the University of Notre Dame said, "I can't think of a morally acceptable reason to clone a human being."

What Father McCormick, in particular, would have us believe is that the ethically correct thing to do in such a painful situation is to let an older child *die* and not have another. He would say that it's better to have zero children to love instead of two.

Bioethicists and others who condemned the Ayalas all felt that they were having a baby for the *wrong reason*. But listen to what Michael Specter of the *Washington Post* wrote in response to this:

> *Can anybody out there provide a universal definition of a good reason to have a baby? Is it better to have a baby because your friends are having a baby, or because your marriage seems like it's missing something, than to have a baby to save a life? What about all the parents who have a second child solely because they don't want an "only child"? Is it better to have a baby to give your first kid a playmate than to have one to save your first kid's life?*

Actually, even in the late twentieth century, millions of people still have babies without any forethought at all. These babies are the unintended by-product of the instinctive urge to relieve sexual tension through the act of intercourse. That's it! Is this biological instinct unethical as well, even within the context of a stable marriage?

If we examine our own experiences, we find that most mothers and fathers give absolute and unqualified love to the children they raise, no matter what reasons were or were not considered in their conception, and no matter where or how this conception occurred. There are exceptions, of course. In some cultures, it is seen as inappropriate for men to involve themselves in the nurturing process. And in all cultures and eras, there will be women who seem to lack a maternal instinct, who do not bond well with the children they bear. You may look at such mothers and wonder why they had children, or why they didn't put them up for adoption. But would you favor state removal of these children from their parents if there is no child abuse? Would you want to outlaw the birth of children to women who test low on the maternal instinct scale? To most Americans, this sort of talk is ridiculous. The right of married adults to have and raise a family, no matter how good they are at it, is constitutionally protected.

Now think about the nurturing instincts of the Ayalas, and the lengths to which they were willing to go to save the life of their older daughter. On what grounds could anyone possibly have imagined that they wouldn't treat a newborn child with the same love and affection?

Indeed, most parents are willing to go to great lengths to protect the lives of their children when confronted with adversity. And although relatively rare, the particular situation in which a couple has another child to act as a potential donor for an older sibling has occurred in hundreds of other families in this country alone, but almost always under the cloak of secrecy, according to informal surveys of organ transplantation centers. And in most of these cases, there was only a 25 percent chance that the newborn child would show compatibility to the older sibling in need of a transplant. Now imagine what these parents would do if cloning became an option. Indeed, ask yourself what you would do in similar circumstances.

Many people will agree that a child in need of a compatible donor provides a compelling, and ethically sound, reason for a small number of parents to engage in cloning. Are there other compelling situations in which the cloning of children might be deemed ethical?

Let's consider the most extreme version of a second scenario. Imagine a young couple that has had a pair of healthy twins—a boy and a girl—the old-fashioned way. Several months after the birth, the mother undergoes chemotherapy as a treatment for cancer. The treatment is a success, but she is now completely sterile. She could have chosen to freeze some of her eggs before her treatment, but she didn't think it was necessary since she didn't intend to have more than two children. And then tragedy strikes. A car driven by an intoxicated man jumps the curb and hits the double baby carriage being pushed by the nanny. Both babies are rushed to the hospital, and both die shortly after arrival.

The pain felt by the parents is unbearable. Not only have they lost their two children, but they believe there is no way they can have any more children that are biologically related to both of them. Unbeknownst to them, however, a young physician in the emergency room at the time the babies were brought in carefully retrieved tissue samples from both bodies shortly after their death, and froze them in a special way. Two years later, as the parents are beginning to accept their sorrow, the doctor reveals to them the existence of the frozen cells. He explains how they might be able to use these samples in an attempt to have their own biological children once again through the process of cloning. Of course, he cautions, they won't be able to bring back their original twins. But the newly born children will look, and most likely act, in a similar way to the six-month-old babies they lost.

The distraught couple is confused by the choice they are being offered. But a genetic counselor helps them understand the process of cloning. And after talking to a sympathetic family minister, they make the decision to go ahead.

The frozen cells are thawed and used to produce embryos. Two em- bryos—one derived from each original twin—are introduced into the mother's uterus. Only one implants, giving rise to a healthy baby girl, nine months later. When this child reaches the age of three, the parents decide to have a second child based on a cell obtained from the other original twin. And a year later, after several failed implantation attempts, a healthy baby boy is born. Their family is now complete and quite ordinary, with a four year old girl and a newborn boy. A stranger would never know that this family was built through the cloning of earlier born fraternal twins. Both children will grow up in a loving environment, and when they are old enough to understand, their parents will explain how they came into existence.

It is hard to imagine what could possibly be wrong with this use of human cloning technology. Indeed, based on a constitutionally protected right to reproduce, it is hard to imagine how it could possibly be ethical to *withhold* the technology—once it is deemed safe—from the couple in this extremely unusual circumstance.

What about another couple whose situation is not so extreme, who become sterile after having one healthy child, and then want to have a second by cloning the first? The second child, of course, would be a later-born identical twin. Would this be unacceptable because the older twin is not in a position to consent to being cloned? I think not. Why is the older child's consent necessary when the parents are simply creating another one of their *own* children with genes that came originally from them! Naturally born identical twins, triplets, and quadruplets, of course, don't consent to the birth of one another.

But there are the new child's emotions to think about. How will she feel when she grows up to find out that she has the same genetic material as her older sister? Will she feel *so* bad that it would have been better for her not to have been born at all? I doubt it. Children born through standard IVF do not seem to be psychologically harmed when they find out they were conceived in a laboratory dish, and if cloning is incorporated as an acceptable practice in the future reprogenetic toolbox, it will be no stranger than IVF once was.

Perhaps there will be a small number of cases where a later-born identical twin is not treated with the respect and dignity that she deserves. Perhaps there will even be child abuse. Yet there is no reason to expect abuse to take place simply because the child is a later-born identical twin. And if it occurs, it is the child abuse we should condemn, no matter what went into the child's creation.

If we accept the use of cloning by sterile parents with one child, do we also accept its use by sterile parents who already have two, three, or four? And what about sterile parents who want to have multiple clones— two, three, or four—from the same older child? What about nonsterile parents who, for some reason, would rather have a later-born identical twin than a nontwin? If you think we should draw a line somewhere, where should it be drawn and who should draw it? And if we draw a line for cloning, why don't we also place a similar limit on the number of children that parents can have the old-fashioned way? Why don't we force women carrying identical twins to abort one so that the remaining child is not subject to the psychological trauma presumed to occur from the realization that he is not genetically unique?

THE CLONING OF ADULTS

Up to 18 percent of all heterosexual couples are infertile from the beginning of their relationship. One hundred percent of all homosexual couples must also be considered infertile, in biological terms, since they cannot reproduce naturally with each other. Before cloning became a possibility, untreatable infertility forced couples to use sperm or egg donations in order to have a child that was biologically related to one parent. But the intrusion of a stranger's genes into a child can be the cause of emotional pain and resentment, especially for the parent whose genes have not been reproduced.

A couple that is infertile at the outset of their relationship obviously wouldn't have any children to clone. Instead, they could clone one of themselves to avoid introducing foreign genes into the family.

Lesbian couples, in particular, would have a new way to share biological parentage of a child. One member of the couple could provide the donor cell, and either one could provide the unfertilized recipient egg. The newly formed embryo could then be introduced into the uterus of the genetically unrelated woman. The child that is born would be related by genes to one mother, and related by birthing to the other, so that both women could rightly call themselves biological parents.

Cloning could be used by infertile heterosexual couples in exactly the same way, as long as the female partner is able to carry a fetus to birth. If the father provided the donor cell to clone from, once again, both members of the couple could rightly call themselves biological parents.

Would a fertile couple have any reason to clone? For the vast majority, the answer would be no. Instead, what most happily bonded couples have always wanted to do—and always will want to do—is to produce a child that represents the ultimate consummation of their love for each other: a child that comes not from one parent or the other, but one that mixes together both their heritages. Some have suggested that cloning could provide a means for preventing the transmission of a deadly disease gene from one parent to his offspring. But this goal can be more easily achieved by selecting embryos—derived from both parents—that don't carry the disease mutation, as we discuss later in this book.

What about single individuals who want to become single parents? Women, in particular, now have the power to do it all by themselves. They can combine one of their skin cells with one of their unfertilized eggs to be placed back into their own uterus to develop into their cloned child. Men, of course, could clone themselves only through the services of a surrogate mother.

How many people would actually want to clone themselves? If the polls are to be believed on this question, the answer is 6 to 7 percent of the American adult population. That's sixty out of every one thousand people questioned, or 5 million or more American adults of reproductive age. This number seems so high as to evoke the suspicion that many who said yes to cloning themselves were not treating the question seriously. Of those who were serious, many would probably change their minds if actually given the chance. Still, there are sure to be people remaining—small in percentage, but large in number—who *would* be willing and ready to clone themselves, if given the chance.

Are all these people egomaniacs, as many have suggested in the media? As a possible answer, consider the fictitious account of Jennifer and Rachel, which begins in the year 2049.

JENNIFER AND RACHEL

Jennifer is a self-sufficient single woman who lives by herself in a stylish apartment on Manhattan's Upper West Side. She has focused almost all her energies on her career since graduating from Columbia University, fourteen years earlier, and has moved steadily upward in the business world. In financial terms, she is now quite well off. In social terms, she is happy being single. Jennifer has had various relationships with men over the years, but none was serious enough to make her consider giving up her single lifestyle.

And then on April 14, 2049, the morning of her thirty-fifth birthday, Jennifer wakes up alone, in her quiet room, before the break of dawn, before her alarm is set to go off, and she begins to wonder. With her new age—thirty-five—bouncing around in her mind, a single thought comes to the fore. "It's getting late." she tells herself.

It is not marriage or a permanent relationship that she feels is missing, it is something else. It is a child. Not any child, but a child of her own to hold and to love, to watch and to nurture. Jennifer knows that she can afford to raise a child by herself, and she also knows that the firm she works for is generous in giving women the flexibility required to maintain both a family and a career. And now she feels, for the first time, that she will soon be too old to begin motherhood.

Jennifer is a decisive woman, and by the end of that day she decides to become a single mother. It is the same positive decision that hundreds of thousands of other woman have made before her. But unlike twentieth-century women, Jennifer knows there is no longer any reason to incorporate a sperm donor into the process. An anonymous sperm cell could introduce all sorts of unknown, undesirable traits into her child, and Jennifer is not one to gamble. Instead, she makes the decision to use one of her own cells to create a new life.

Jennifer is well aware that federal law makes cloning illegal in the United States except in cases of untreatable infertility. She realizes that she could get around the law through a marriage of convenience with a gay friend, who would then be declared infertile by a sympathetic physician. But she decides to do what increasing numbers of other women in her situation have done recently—take an extended vacation in the Cayman Islands.

On Grand Cayman Island, there is a large reprogenetic clinic that specializes in cloning. The young physicians and biologists who work at this clinic do not ask questions of their clients. They will retrieve cells from any willing adult, prepare those cells for fusion to unfertilized eggs recovered from any willing woman, and then introduce the embryos that develop successfully into the uterus of the same, or another, willing woman. The cost of the procedure is $80,000 for the initial cell cloning and embryo transfer, and $20,000 for each subsequent attempt at pregnancy if earlier embryos fail to implant. When the clinic first opened, the fees were twice as high, but they dropped in response to competition from newly opened clinics in Jamaica and Grenada.

Since Jennifer is a healthy fertile woman, she has no need for other biological participants in the cloning process. A dozen unfertilized eggs are recovered from her ovaries and made nuclear-free. One-by-one, each is fused with a donor cell obtained from the inside of her mouth. After a period of incubation, healthy-looking embryos are observed under a microscope, and two of these are introduced into her uterus at the proper time of her menstrual cycle. (The introduction of two embryos increases the probability of a successful implantation.) After the procedure, Jennifer stays on the island three more days to rest, then flies back to New York.

A week later, Jennifer is thrilled by the positive blue + symbol that appears on her home pregnancy test. She waits another two weeks to confirm that the pregnancy has taken with another test, and then schedules an appointment with Dr. Steve Glassman, her gynecologist and obstetrician. Dr. Glassman knows that Jennifer is a single woman, and he doesn't ask—and Jennifer doesn't tell—how her pregnancy began. The following eight and half months pass by uneventfully, with monthly, then weekly, visits to the doctor's office. Ultrasound indicates the presence of a single normal fetus, and amniocentesis confirms the absence of any known genetic problem. Finally, on March 15, 2050, a baby girl is born. Jennifer names her Rachel. To the nurses and doctors who work in the delivery room, Rachel is one more newborn baby, just like all the other newborn babies they've seen in their lives.

Jennifer, holding Rachel in her arms, is taken to a room in the maternity ward, and shortly thereafter, the nurse on duty brings by the form to fill out for the birth certificate. Without a word, she enters Jennifer's name into the space for "the mother." She then asks Jennifer for the name of the father. "Unknown," Jennifer replies, and this is duly recorded. A day later, Jennifer is released from the hospital with her new baby girl.

Rachel will grow up in the same way as all other children her age. Occasionally, people will comment on the striking similarity that exists between the child and her mother. Jennifer will smile at them and say, "Yes. She does have my facial features." And she'll leave it at that.

From time to time, Jennifer will let Rachel know that she is a "special" child, without going into further detail. Then one day, when her daughter has grown old enough to understand, Jennifer will reveal the truth. And just like other children conceived with the help of reprogenetic protocols, Rachel will feel . . . special. Some day in the more distant future, when cloning becomes just another means of alternative reproduction, accepted by society, the need for secrecy will disappear.

Who is Rachel, and who really are her parents? There is no question that Jennifer is Rachel's birth mother, since Rachel was born out of her body. But, Jennifer is not Rachel's genetic mother, based on the traditional meanings of mother and father. In genetic terms, Jennifer and Rachel are twin sisters. As a result, Rachel will constantly behold a glimpse of her future simply by looking at her mother's photo album and her mother herself. She will also understand that her single set of grandparents are actually her genetic parents as well. And when Rachel grows up and has children of her own, her children will also be her mother's children. Thus, with a single act of cloning, we are forced to reconsider the meaning of parents, children, and siblings, and how they relate to one another.

IS CLONING WRONG?

Is there anything wrong with what Jennifer has done? The most logical way to approach this question is through a consideration of whether anyone, or anything, has been harmed by the birth of Rachel? Clearly no harm has been done to Jennifer. She got the baby girl she wanted and she will raise her with the same sorts of hopes and aspirations that most normal parents have for their children.

But what about Rachel? Has she been harmed in some way so detrimental that it would have been better had she not been born? Daniel Callahan, the Director of the Hastings Center (a bioethics think tank near New York City), argues that "engineering someone's entire genetic makeup would compromise his or her right to a unique identity." But no such "right" has been granted by nature—identical twins are born every day as natural clones of each other. Dr. Callahan would have to concede this fact, but he might still argue that just because twins occur naturally does not mean we should create them on purpose.

Dr. Callahan might argue that Rachel is harmed by the knowledge of her future condition. He might say that it is unfair for Rachel to go through her childhood knowing what she will look like as an adult, or being forced to consider future medical ailments that might befall her. But even in the absence of cloning, many children have some sense of the future possibilities encoded in the genes they got from their parents. In my own case, I knew as a teenager that I had a good chance of inheriting the pattern baldness that my maternal grandfather expressed so thoroughly. Furthermore, genetic screening already provides people with the ability to learn about hundreds of disease predispositions. And as genetic knowledge and technology become more and more sophisticated, it will become possible for any human being to learn even more about their genetic future than Rachel can learn from Jennifer's past. In American society, it is generally accepted that parents are ultimately responsible for deciding what their children should, or should not, be exposed to. And there's no reason to expect that someone like Jennifer would tell Rachel something that was not in her best interest to know.

Just because Rachel has the same genes as Jennifer does not mean that her life will turn out the same way. On the contrary, Rachel is sure to have a different upbringing in a world that has changed significantly since her mother's time. And there is no reason why she can't chart her own unique path through life. Furthermore, when it comes to genetic predispositions, they are just that and nothing more. Although their genetically determined inclinations may be the same, mother and daughter may choose to follow those inclinations in different ways, or not at all.

It might also be argued that Rachel is harmed by having to live up to the unrealistic expectations that her mother will place on her. But there is no reason to believe that Jennifer's expectations will be any more unreasonable than those of many other parents who expect their children to accomplish in their lives what the parents were unable to accomplish in their own. No one would argue that parents with such tendencies should be prohibited from having children. Besides, there's no reason to assume that Jennifer's expectations will be unreasonable. Indeed, there is every reason to believe Rachel will be loved by her mother no matter what she chooses to do, as most mothers love their children.

But let's grant that among the many Rachels brought into this world, some *will* feel bad that their genetic constitution is not unique. Is this alone a strong enough reason to ban the practice of cloning? Before answering this question, ask yourself another: Is a child having knowledge of an older twin worse off than a child born into poverty? If we ban the former, shouldn't we ban the latter? Why is it that so many politicians seem to care so much about cloning but so little about the welfare of children in general?

Some object to cloning because of the process that it entails. The view of the Vatican, in particular, is that human embryos should be treated like human beings and should not be tampered with in any way. However, the cloning protocol does *not* tamper with embryos, it tampers only with *unfertilized* eggs and adult cells like those we scratch off our arms without a second thought. Only after the fact does an embryo emerge (which could be treated with the utmost respect if one so chooses).

There is a sense among some who are religious that cloning leaves God out of the process of human creation, and that man is venturing into places he does not belong. This same concern has been, and will continue to be, raised as each new reprogenetic technology is incorporated into our culture, from in vitro fertilization twenty years ago to genetic engineering of embryos—sure to happen in the near future. It is impossible to counter this theological claim with scientific arguments. We will come back to God's domain in the last part of this book.

Finally, there are those who argue against cloning based on the perception that it will harm society at large in some way. The *New York Times* columnist William Safire expresses the opinion of many others when he says, "cloning's identicality would restrict evolution." This is bad, he argues, because "the continued interplay of genes . . . is central to humankind's progress." But Mr. Safire is wrong on both practical and theoretical grounds. On practical grounds, even if human cloning became efficient, legal, and popular among those in the moneyed classes (which is itself highly unlikely), it would still only account for a fraction of a percent of all the children born onto this earth. Furthermore, each of the children born by cloning to different families would be different from one another, so where does the identicality come from?

On theoretical grounds, Safire is wrong because humankind's progress has nothing to do with unfettered evolution, which is always unpredictable and not necessarily upward bound. H. G. Wells recognized this principle in his 1895 novel *The Time Machine,* which portrays the natural evolution of humankind into weak and dimwitted, but cuddly little creatures. And Kurt Vonnegut follows this same theme in *Galápagos,* where he suggests that our "big brains" will be the cause of our downfall, and future humans with smaller brains and powerful flippers will be the only remnants of a once great species, a million years hence.

Although most politicians professed outrage at the prospect of human cloning when Dolly was first announced, Senator Tom Harkin of Iowa was the one lone voice in opposition. "What nonsense, what utter utter nonsense, to think that we can hold up our hands and say, 'Stop,' " Mr. Harkin said. "Human cloning will take place, and it will take place in my lifetime. I don't fear it at all. I welcome it."

As the story of Jennifer and Rachel is meant to suggest, those who want to clone themselves or their children will not be impeded by governmental laws or regulations. The marketplace—not government or society—will control cloning. And if cloning is banned in one place, it will be made available somewhere else—perhaps on an underdeveloped island country happy to receive the tax revenue. Indeed, within two weeks of Dolly's announcement, a group of investors formed a Bahamas-based company called Clonaid (under the direction of a French scientist named Dr. Brigitte Boisselier) with the intention of building a clinic where cloning services would be offered to individuals for a fee of $200,000. According to the description provided on their web page (http://www.clonaid.com), they plan to offer "a fantastic opportunity to parents with fertility problems or homosexual couples to have a child cloned from one of them."

Irrespective of whether this particular venture actually succeeds, others will certainly follow. For in the end, international borders can do little to impede the reproductive practices of couples and individuals.

SURREPTITIOUS CLONING

In democratic societies, people have the right to reproduce and the right to *not* reproduce. This last "right" means that men and women cannot be forced to conceive a child against their will. Until now, it has been possible to exercise this particular right by choosing not to engage in sexual intercourse, and not to provide sperm or eggs for use in artificial insemination or IVF. But suddenly, human cloning opens up frightening new vistas in the realm of reproductive choice, or lack thereof. Suddenly, it becomes possible to use the genetic material of others without their knowledge or consent.

Let's reconsider the Jennifer and Rachel scenario in the light of reproductive choice. At first glance, it might seem that nothing is amiss here because Jennifer obviously gave her consent to be cloned. But reproductive choice has been interpreted traditionally to mean that people have the right not to be genetic parents against their will. Does this mean that Jennifer should have asked her own parents for permission to create a clone—her identical twin and their child—before proceeding? Actually, all of *your* genes, as well, came from *your* mother and father. Does this mean that your parents have the right to tell you how to use them?

At least Jennifer gave her consent to be cloned. But what are we to make of a situation in which someone is cloned without his or her knowledge, let alone consent. It takes only a single living cell to start the cloning procedure, and that cell can probably be obtained from almost any living part of the human body. There are various ways in which cells could be stolen from a person. I will illustrate one here with what I will call the Michael Jordan scenario.

Let's move to the near future. The year is 2009, and Jordan has now retired as a professional basketball player. He goes into his doctor's office for his annual checkup, during which a blood sample is taken into a standard tube. Jordan's sample, along with others, is given over to a medical technician, who has been waiting for this moment since Jordan scheduled his appointment a month before. After closing the lab door behind her, she opens the tube of Jordan's blood and removes a tiny portion, which is transferred to a fresh tube that is quickly hidden in her pocket. The original tube is resealed, and no one will ever know that it has been tampered with.

At the start of her lunch break, the technician rushes the tube of blood to her friend at a private IVF clinic on the other side of town. The small sample is emptied into a laboratory dish, and there Jordan's white blood cells are bathed in nutrients and factors that will allow them to grow and multiply into millions of identical cells, each one ready for cloning. The cells are divided up into many portions, which are frozen in individual tubes for later use.

And then the word goes out on the street. For a $200,000 fee, you can have your very own Michael Jordan child. Would anyone buy? If not a Michael Jordan child, would they be interested in a Tom Cruise, a Bill Clinton, or a Madonna (the singer not the saint)?

It's important to understand that what most people want more than anything else is to have their *own* child, not someone else's child, no matter who that someone else might be. And if cloning someone else is an option, then cloning oneself is also an option. So what possible reason could exist for choosing a genetically unrelated child?

Perhaps heartless mothers will want a clone of someone famous in the belief that they will prosper on the income that a clone could make, or the fame that he would bring. But it would require an enormous investment in time and money to raise a child over many years before there was even a chance of a payback. Clones of Michael Jordan would likely be born with the potential to become outstanding athletes, and clones of Tom Cruise or Madonna might have the same artistic talent as their progenitors. But the original Jordan, Cruise, and Madonna owe their success even more to hard work than genetic potential.

Clones might not have the same incentive to train and exert themselves even if—and perhaps because—unscrupulous parents and promoters try to force them in a specified direction against their will. And while one Madonna clone might attract fame and attention, the next dozen will almost certainly be ignored. It is hard to imagine that many potential parents would be willing to take this gamble, with the wait being so long, and the chances of success so small.

There will probably always be some infertile couples or individuals who will want to clone simply for the opportunity to raise a child who is likely to be beautiful or bright, without any desire to profit from the situation themselves. These people will be able to reach their reproductive goals by cloning someone—who is not famous—*with* their consent. In the future, cell donors could be chosen from a catalog in the same way that sperm and egg donors are chosen today (as described in chapter 13).

In contrast, cloning surreptitiously will almost certainly be frowned upon even by those who accept other uses of the cloning technology. And those who participate will run the risk of serious litigation on the basis of infringing upon someone else's reproductive rights. This is not to say, however, that surreptitious cloning will never occur. On the contrary, if something becomes possible in our brave new reproductive world, someone will probably do it, somewhere, sometime.

10

Where Will Cloning Lead Us?

The focus of the chapter you have just read—and the public mind in general—has been on the use of cloning to reproduce children who are later-born identical twins of people already alive. It is this idea that scares people and causes politicians to promote laws for banning all uses of the technology. But what has not come forth as clearly are the ways in which cloning can be used in combination with other reprogenetic technologies to solve a whole range of biomedical problems *without* producing later-born twins.

Indeed, there are many biomedical scientists who believe that the real significance of cloning lies not in what the technology can do by itself, but in its enabling contribution to other areas of reprogenetics. There are two fields of research, in particular, that will gain substantially from the incorporation of cloning. These are tissue regeneration and genetic engineering.

TISSUE REGENERATION

Let's start by returning to the story of Anissa Ayala. Anissa's own bone marrow was contaminated with cancerous cells. There was no way to kill just the cancerous cells and leave the healthy ones behind; Anissa's cancer could only be purged through the complete eradication of all bone marrow cells. But bone marrow cells are required to replenish blood cells. So Anissa would have died unless she received a bone marrow transplant from a compatible donor. As noted, the only perfect donor is an identical twin. And if a future Anissa is not blessed with an identical twin from birth, cloning could provide her family with one later on.

Unfortunately, there will be many situations in which a child will be in need and cloning will not be feasible or acceptable. Some families will be opposed on moral grounds. Others will not be in a position to bring a new baby into their lives, because they may be too old or just not able to raise another child. Fortunately, cloning technology will provide the basis for an alternative solution.

The problem faced by Anissa and all other people who have cancerous or failing organs or tissues is that once a person is born, his or her body no longer has a reservoir of undifferentiated cells that might be coaxed into recreating a new version of the defective tissue. The organs and tissues present in a child or adult have reached their final stage of differentiation and can't be changed into each other. Skin cells can't be converted into bone marrow cells, and blood cells can't be converted into liver cells.

In contrast, the cells of the early embryo have the capacity to turn into each of these adult cell types, as well as every other cell present in every tissue and organ. If scientists could figure out how embryonic cells turn into a particular tissue, they might be able to force a cloned embryo along just this pathway, bypassing the need to create a new human life. In fact, scientists were working on this very problem for almost two decades before cloning became a reality.

Normally, when an embryo grows—by producing more cells—it also *develops,* and it is through the process of development that cells differentiate into various kinds of tissues. The pathway of development is rather strict, and on each day following conception, the embryo, and then the fetus, assume a well-defined shape and structure. If you are pregnant and you know the approximate date on which conception took place, you can look at the color pictures in the 1990 book by Lennart Nilsson called *A*

Child Is Born to see what the embryo or fetus inside you looks like, and with each passing week, you can watch it change in form as it develops.

The critical point is that growth and development are coupled. You don't get one without the other. Not until 1981, that is, when embryologists in the United States and England succeeded in perfecting methods to get embryonic growth, *without* development, in a laboratory dish. They accomplished this feat by tricking the embryonic cells into "thinking" they were still present in a very young embryo at a stage where division is *supposed* to occur without differentiation. This deception is carried out by placing the embryo in an environment overloaded with early embryo molecular signals. In this environment, an embryo will continue to grow and divide, over and over again, to produce millions upon millions of identical cells that are all frozen—in a developmental sense—at the same early embryonic stage. Scientists refer to these cells as embryonic stem cells, or just ES cells for short.

What the ES cell technology provides is a tool for expanding the embryo into a mass of undifferentiated tissue of any size that is needed. After this first step is accomplished, it then becomes possible to convert this undifferentiated mass into the particular tissue that one desires. Once again, the feat is accomplished with the use of particular molecular signals.

Just as certain signals can be used to fool ES cells into remaining in an embryonic state, other signals can be used to force them down specific pathways of differentiation in a controlled manner. Some signals might be used to turn them into bone marrow cells and others could be used to turn them into primitive nerve cells, for example. And what will almost certainly happen over the next twenty years is that scientists will discover what signals are needed to convert embryonic cells into every tissue that exists in the adult human body.

Now let's return to our future Anissa, in need of a bone marrow transplant, and outline the steps we might be able to take to provide it to her. First, cloning technology would be used to convert one of her skin cells into a brand-new embryo, with the help of a donated unfertilized egg. But rather than allow this embryo to develop into a fetus, scientists would use special molecules to expand it into a large mass of embryo-like ES cells. Once a sufficient number of cells had been obtained, they would be incubated with a different set of molecular signals to convert them all directly into bone marrow cells, which could be used for transplantation back into the body from which they originally came.

The same basic technology may be used—with different signals—to cure a host of other diseases. Nerve cells might be generated as a cure for Parkinson disease, newly minted heart or liver cells might augment the function of other ailing organs, and freshly grown blood vessels could be used to replace those that are damaged by arteriosclerosis. And in each of these cases, cell or tissue replacement therapy could be accomplished in a few weeks with a high degree of success and efficiency, and without any of the dilemmas that might arise in cases where it is necessary to bring another child into the world.

GENETIC ENGINEERING

There is a final consequence of cloning that is more significant and powerful than any other use of the technology, one that has the potential to change humankind: the genetic engineering of human beings. Without cloning, genetic engineering is simply science fiction. But with cloning, genetic engineering moves into the realm of reality.

First, a quick definition. When I use the term "genetic engineering," it will refer to the process by which scientists alter or add specific genes to the genetic material present in the embryo so that an individual could be born with characteristics that he or she would not have had otherwise. Cloning, by itself, is not genetic engineering.

Since the 1980s, genetic engineering has been practiced with success in animals like mice, cows, sheep, and pigs. But it has yet to be applied to human beings for one simple reason—it is incredibly inefficient. With the simplest technique for adding genes to embryos, the success rate is 50 percent at best, and this is accompanied by a 5 percent risk of inducing disease-causing mutations in the animal that is born. That's not a problem for animal geneticists—who can choose the one healthy animal with a desired genetic modification from among a litter or flock—but it is unacceptable for use with humans. And with more sophisticated techniques of gene alteration, the problem just gets worse, with only one cell in a million likely to be altered in the correct way.

With such a low rate of success, the direct engineering of genes within an isolated human embryo—destined to be a child—is not something that anyone would try or accept. But with cloning, the entire equation changes. Now, multiple cells grown from a single embryo could be subjected to genetic engineering. With protocols already available today, those that appear to be engineered as desired could be recognized and picked out. Each single selected cell could be expanded by itself into a clone of cells that provides sufficient material for the confirmation of genetic integrity. Then, and only then, would one cell from this mass of cells be used by means of nuclear transplantation to produce a new embryo, which would develop into a new human being, with a special genetic gift. Incredibly, within five months of the announcement of Dolly's birth, on July 25, 1997, the same team of Scottish scientists announced that they had successfully carried out this very protocol with the birth of several lambs carrying a foreign human gene. It is in the very same manner—when the techniques of cloning and genetic engineering are combined—that the human species will gain control over its own destiny, as we will explore later in this book.

PART FOUR

MOTHERS AND FATHERS: VARIATIONS ON A THEME

And when Rachel saw that she bore
Jacob no children, . . . she said to Jacob,
"Give me children, or I shall die!"

—GENESIS 30:1

11

Three Mothers and
Two Fathers

In most societies, at most times, most children have had a single mother and a single father. This is the norm for our species. But as long as our species has existed, there have been exceptions. And though exceptions are quite common today, the language we use to distinguish among the different types of parents that a child can have is still less than ideal.

Unfortunately, it is difficult to find words that are not emotionally laden. When one woman gives birth to a child that is raised by another, who is the "real mom"? From a biological perspective, it would be the woman who gave birth, but from a social perspective, it would be the woman who raised the child. Confusion occurs as a result of the same problem that we encountered in the meaning of the word *life* itself. Just as that single word is used to describe life at both the cellular level and the conscious level, so the words *mother, father,* and *parent* can be used to describe individuals who make a biological contribution, as well as those who contribute socially. In the most global definition of the word, all mothers are "real," they are just different.

I will use *genetic father* or *bio-dad* to describe the man who contributes a sperm nucleus toward the creation of a child. Until twenty years ago, it was also possible to speak about a *biological mother* or *bio-mom* in an unambiguous way. But with the advent of IVF and embryo transfer, the two essential woman-contributed biological ingredients can be separated so that a child can now have two bio-moms. Whenever it is necessary to distinguish between them, I will use *genetic mother* or *gene-mom* to describe the woman who contributed the egg, and *gestational mother* or *birth-mom* to describe the woman whose womb the fetus developed in.

The terms used most frequently to distinguish the parents who raise a child from the biological parents are *rearing* mothers and fathers or *social* mothers and fathers. In the case of adoption, the terms *adoptive* mother and father are applied. Occasionally, the term *nurturing mother* is also used, but this is ambiguous since *nurturing* can be interpreted as describing a temperament rather than a parental role. Most of the time, the parents who raise a child are referred to simply as the mother and father. It is only when biological parents are discussed that it becomes necessary to make a distinction. And when this distinction needs to be made, I will use *social mother* and *social father* for the lack of any better terminology.

Another way in which mothers and fathers are distinguished is through the law. But there is no absolute correspondence between a *legal mother* or a *legal father* and either biological or social parents. In cases where two mothers or fathers exist, the *legal* designation can be applied to either one depending on the laws of the state or the ruling of a court. Furthermore, the designation can change in response to further court rulings. Nevertheless, legal mothers and fathers have both rights and responsibilities toward their children that are strictly defined.

To recapitulate, there are two possible types of fathers—bio-dads and social dads. And there are three types of mothers—gene-moms, birth-moms, and social moms. Obviously, the bio-dad and the social-dad can be the same person, as can the three different types of moms, which is typically the case. But as we shall see shortly, it is also possible for any two of the three mother types to be combined into one woman, while a second woman takes on the role of the third mother type.

ONE SOCIAL MOTHER AND ONE SOCIAL FATHER, MORE OR LESS

Before we delve into more complicated possibilities, it is important to consider the most common exception to the traditional family model of a single bio-social father and single bio-social mother: the single-parent family. Throughout history, single bio-moms have raised their children alone because bio-dads died, took off, or were thrown out. In modern times, single-parent families centered around a bio-dad have also become commonplace.

In addition to single-parent families formed by happenstance, there are now individual adults who consciously make the choice to become single bio-parents so they can raise their children without interference from a spouse or partner. For women, this is easy to accomplish. In fact, a book entitled *Having a Baby Without a Man: The Woman's Guide to Alternative Insemination* was co-authored by a female physician to help women consider obstacles they might encounter as they proceed along this path. Biologically speaking, the only thing that a woman needs *today* is a willing male to participate in intercourse or provide sperm for artificial insemination. If she desires an anonymous source of genetic material, she can go to a sperm bank. In the future, when cloning becomes available, a man won't be required at all.

For men, the task is more complicated but not impossible. A publication on the World Wide Web called *Fathering Magazine* (http://www.father mag.com) provides a virtual support group for such men as well as "involved fathers, fathers who are the primary parent, single fathers, and men aspiring to become fathers." In an interactive forum on this "webzine" one man wrote:

> It is nice to know that there are other men who, like myself, have intentionally become a single parent. LOTS [original emphasis] of women have done it, and I'm sure some other men have, and others would like to. After all the disagreement I went through with my wife over the raising of my first two children, I resolved to do it again the way I wanted. It took me a couple of years to get the arrangement I wanted, but it was worth it, and I'm very pleased with the results so far. One way to get around the way the system favors mothers is to

just exclude them. At least we dads have the encouraging single-father statistics about how the kids turn out.

While the single-parent family excludes a social parent of one sex, another class of nontraditional families embraces a social parent of the same sex. Children in these families have two social mothers or two social fathers who act as a married couple, in effect if not in law. One recent survey suggests that up to 6 million children may be living in such families.

Until recently, it was difficult for both members of a same-sex couple to be recognized as the joint social parents of a child. The problem was that antiquated laws allowed recognition of only a single legal mother and a single legal father. As a consequence, a child could only be adopted by the female partner of a bio-mom if the parental rights and responsibilities of the bio-mom herself were terminated. But over the last decade, gay couples have been granted what are commonly called "second parent adoptions" in fifteen states. And once an adoption is granted, the Constitution of the United States guarantees legal recognition of these joint parenting arrangements in all other states of the Union.

WHEN A BIOLOGICAL INGREDIENT IS MISSING

Until the advent of cloning, there were three distinct biological ingredients that were required to bring forth each newborn baby into the world—an egg, a sperm nucleus, and a womb. When one of these ingredients was missing, the result was infertility. As we discussed earlier, medical science has developed increasingly sophisticated treatments, culminating in the use of IVF with other protocols, to help infertile couples give birth to biological children. But these medical treatments are far from a panacea. First, there are some forms of infertility that are too severe to be treated even with current IVF technology. Second, more than half of all couples who enter an IVF program fail in their effort to obtain biological children. And finally, the very high cost and low success rate of IVF stops many couples from pursuing this possibility in the first place.

When the desire to have and raise children is intense, and when all other options for curing infertility have been exhausted, many couples are forced to consider other approaches to building a family. By necessity, all of these approaches are dependent upon the inclusion of one or more biological parents from outside the social family unit.

The least technical choice is simple adoption. With adoption, all of the biological ingredients of reproduction are contributed by a man and woman other than the social father and mother. As a consequence, there is a complete division between a child's biological and social heritage. While adoptions will always occur, the number of healthy newborn babies that are made available each year in the United States is far lower than it was in the past. This is owing in large part to the 1973 U. S. Supreme Court decision to legalize abortion, which brought about a dramatic reduction in the number of unwanted pregnancies that go full-term. Also, the general societal acceptance of premarital sex has reduced the stigma of being an unwed mother. As a consequence, young unmarried women who typically gave up their children for adoption in earlier times are now being encouraged (by social workers) to take the opposite course. The result of these converging factors is that in 1984, 2 million infertile American couples competed for 58,000 newborn American children.

Aside from adoption, other traditional approaches to overcoming infertility are based on methods of "collaborative reproduction," to use a term coined by the bioethicist John Robertson. Collaborative reproduction is performed by combining biological ingredients from one or both prospective social parents with ingredients "donated" by one or two nonsocial biological parents.

Collaborative methods exist for restoring any one or two of the three biological ingredients that are required to create a child. Egg and womb are restored through the use of a surrogate mother who is inseminated with the prospective father's sperm. The sperm alone is restored by artificial insemination of the prospective mother with a donated sample. The egg alone is restored by IVF with a donated egg and the prospective father's sperm, followed by the introduction of the embryo into the prospective mother's uterus. The egg and sperm together are restored by the introduction of a donated embryo into the prospective mother's uterus. Finally, the womb alone is restored by IVF with sperm and egg both obtained from the prospective parents, followed by the introduction of the embryo into a surrogate mother. Each of these approaches to collaborative reproduction will be explored in the following two chapters.

12

Contracting for a Biological Mother

SURROGACY: OLD AND NEW

"And when Rachel saw that she bore Jacob [her husband] no children, Rachel envied her sister [who had children earlier by Jacob], and said to Jacob, 'Give me children, or I shall die!' And Jacob's anger was aroused against Rachel, and he said, 'Am I in the place of God, who has withheld from you the fruit of the womb?' So she said, 'Here is my maid Bilhah; go in to her, and she will bear a child on my knees, that *I also may have children by her.*' Then she gave him Bilhah her maid as wife, and Jacob went in to her. And Bilhah conceived, and bore Jacob a son. Then Rachel said, 'God has judged my case; and He has also heard my voice and given me a son.' Therefore called she his name Dan."

As this Bible story tells us, surrogacy has been practiced at least as long as there have been historians to record it. Not only Rachel, but Sarah and Leah as well, all directed their Biblical husbands to make their hand-maids pregnant so that they might "also have children by her," as Rachel put it. It is likely that surrogacy has occurred in secrecy throughout the intervening years as well, when infertile women, on occasion, persuaded sisters or friends to have a baby with their husbands that they could then raise.

According to the *Oxford English Dictionary*, the term *surrogate* is defined generally as, "A person . . . that acts for or takes the place of another; a substitute." A 1978 *Time* magazine article provided the first use in the popular media of the term *surrogate mother* to describe a woman who brought a child to birth for another couple to raise as their own. But as soon as the term was coined, it came under attack from ethicists, feminists, and legal commentators who considered the term "inappropriate," "bizarre," and "troublesome." How could the woman who gave birth to a child be the surrogate mother when "she is the *actual* mother"?, they asked.

This is a clear example of how the multiple definitions of the word *mother* can cause confusion. Based on the biological definition, the so-called surrogate is indeed the *real* mother, not a surrogate. But in the eyes of the prospective social mother, the pregnant woman is indeed a biological surrogate.

The pronouncements of academics and other critics have had no effect on the public at large, and *surrogate mother* has become a commonly used and commonly understood term, which has been modified further to describe the two possible ways in which the fetus carried by the surrogate mother could be related to her and the prospective social parents. In "traditional surrogacy," the surrogate mother is both the gene-mom and the birth-mom. This type of surrogacy is typically initiated by artificial insemination with sperm from the prospective social father. However, as the Bible tells us, sexual intercourse between the prospective father and the surrogate mother can also be used in place of artificial insemination.

In "gestational surrogacy," the surrogate mother is only the birth-mom and does not contribute genetic material to the fetus that she carries. This type of surrogacy is typically initiated with the use of gametes from both the prospective mother and prospective father to obtain fertilized eggs by IVF. The embryos that develop are introduced into the surrogate mother's uterus with the hope that one or more will implant and develop.

The modern era of surrogacy by contract did not begin until the late 1970s, and the concept of surrogate motherhood did not enter the public consciousness until the early 1980s, after a series of highly publicized cases of surrogacy arrangements gone awry. Over the following decade, hundreds of articles on surrogacy appeared in popular magazines, academic journals, and books. Feminists, lawyers, ethicists, and theologians all argued the case for or against the practice. In addition, various state courts and legislatures made determinations on the validity of surrogacy contracts and the proper means to resolve disputes when multiple parents want the child.

HOW IT WORKS

When a couple or individual decide to have a child through a surrogate mother, how do they go about doing it? One way is to find a friend or relative who wishes to help them overcome their infertility, or in the case of a single man, an inability to provide the maternal components of the reproductive process. The friend or relative agrees to become pregnant and give the child up upon birth to the prospective parent or parents.

This form of surrogacy is considered "altruistic," since the surrogate is acting primarily to help someone she knows rather than for financial gain. The prospective parents may reimburse the surrogate for medical expenses and perhaps living expenses as well. The individuals involved may or may not decide to formalize their agreement in terms of a legally binding contract.

Becoming a surrogate mother entails an enormous physical and emotional commitment that most women don't want to make, even for a close friend or relative. Thus, more often than not, prospective parents will be unable to find someone to act as an altruistic surrogate for them, and they will have to turn to a commercial surrogacy agency or broker.

Commercial surrogacy costs money, a lot of money. The agency or broker will charge a fee of up to $16,000 for initiating the process, recruiting and evaluating surrogates, and establishing a contract, and the surrogate herself can receive up to $15,000 for the service she performs. Separate reimbursement for medical expenses associated with the pregnancy and birth will add another $5,000. In addition, there will be miscellaneous expenses—for maternity clothes, lost wages near the end of the pregnancy, life insurance, counseling, and legal fees. Total costs can run up to $50,000 for traditional surrogacy arrangements, and more for gestational surrogacy arrangements that involve IVF.

The surrogacy contract—signed by the surrogate mother and the prospective parents—is critical to the process. The contract obviously requires the transfer of the newborn child to the custody of the contracting parents, and almost always requires the birth mother to relinquish any claims of legal ties to the child. A surrogacy contract can also impose other constraints on the behavior of the surrogate woman during her pregnancy. Smoking, drinking, and drug taking are typically forbidden, and specific nutrition as well as a program of prenatal medical care are typically prescribed. Although these constraints are meant to decrease the risk of harm to the developing fetus, some surrogate mothers may feel that they are giving up individual liberty to abide by these rules.

HOW IT'S VIEWED

Many ethicists, lawyers, theologians, and feminists have condemned the practice of surrogacy. They view it as immoral and harmful to the woman who acts as a surrogate, as well as society in general. Many also see harm befalling the baby that is handed over from the surrogate to the contracting couple.

The claim of immorality is made most forcefully by those who view any act of procreation that extends outside the traditional boundaries of sexual intercourse between a married man and woman as a challenge to the will of God. Some have suggested further that certain forms of surrogacy are equivalent to adultery or incest (when, for example, a mother or sister acts as a surrogate mother), even when fertilization is initiated by artificial insemination. This objection results from an inability, or unwillingness, to separate sex and reproduction.

Other critics are most disturbed by the potentially harmful effects that surrogacy can have on the contracting woman. They view the experience as "dehumanizing" to the women involved, and as a consequence, demeaning to society as a whole. Some liken commercial surrogacy to prostitution, seeing an analogy between the rental of a uterus and the rental of a vagina. This link is based on the sense that many woman enter surrogacy contracts out of a desperate need for money. As a result, there is the fear that such women will be exploited by the couples they are contracted to.

Finally, some critics believe that the payment of money to a surrogate mother takes something away from the "human worth" of the child that is born. "What is fundamentally unethical about surrogate mother arrangements is that they . . . treat a person (the child) as though he were a thing, a commodity," says law professor Herbert Krimmel. "Surrogate mother arrangements . . . encourage and tempt the adopting parents to view children as items of manufacture."

I do not believe that arguments along this line, in particular, will ring true for anyone who has lovingly raised a child, and there is no evidence to support Krimmel's claim that social parents view surrogate-borne children as "items of manufacture." On the contrary, studies carried out on families with adopted children indicate that adoptive parents "feel no less bonded to their children than responsible genetic parents."

There are a few prominent ethicist-lawyers, such as John Robertson and Lori Andrews, who support the right of women and men to engage in surrogacy contracts. Andrews, a professor at the Chicago-Kent College of Law, writes, "Symbolic arguments and pejorative language seem to make up the bulk of the policy arguments and media commentary against surrogacy." She goes on to argue that much of the opposition is based on the notion that a woman must be protected from her own inability to act in an appropriate societal manner when confronted with the decision to become a surrogate mother. "But," she points out, "arguing for a ban on surrogacy seems to concede that the *government* [original emphasis], rather than the individual woman, should determine what risk a woman should be allowed to face." She continues, "The rationales for such a ban are often the very rationales that feminists have fought *against* [my emphasis] in the contexts of abortion, contraception, nontraditional families, and employment," and the "rationales . . . to justify this governmental intrusion into reproductive choice may come back to haunt feminists in other areas of procreative policy and family law."

As one surrogate mother, Donna Regan, testified before the New York legislature, "I find it extremely insulting that there are people saying that, as a woman, I cannot make an informed choice about a pregnancy that I carry. [Like everyone else, I] make other difficult choices in my life."

WHEN THINGS GO RIGHT: THROUGH THE LOVE OF TWO MOTHERS

One story of altruistic surrogacy devoid of the kinds of problems that its detractors fear is that of Karen Ferreira-Jorge and her mother Patricia

(Pat) Anthony of the citrus-farming town of Tzaneen, 220 miles north of Johannesburg, South Africa. Karen and her husband Alcino had gotten married with plans to have a large family. But in 1984, during the birth of their first child, Karen almost bled to death from medical complications, and her uterus was removed. Karen and Alcino wanted to have more children, and they considered the possibility of enlisting the services of an unrelated surrogate mother. But they were discouraged by fears that a surrogate mother might renege on her promise to give the baby up after birth.

While Karen and Alcino desperately tried to devise a plan for achieving their reproductive goals, Karen's soon-to-be-forty-eight-year-old mother Pat decided to offer her own services as a gestational surrogate. Pat's husband and Karen's father Raymond embraced the plan, along with Karen and Alcino themselves.

And so in January 1987, Karen underwent ovarian stimulation, resulting in the production of eleven eggs that were fertilized by her husband's sperm. Two days later, four young embryos were introduced into Pat's uterus. Amazingly, given Pat's age, three of the four embryos implanted successfully and began to develop. At dawn on Thursday, October 1, 1987, three healthy babies named David, Jose, and Paula were born by Caesarean section at Park Lane Clinic in Johannesburg. Karen, who had received hormones to stimulate milk production, began to breast-feed all three children within hours of their birth.

This was one surrogacy arrangement in which everyone came out a winner. Karen and Alcino got the large family they had dreamed of. Pat and Raymond got three new grandchildren to spoil in the way that most grandparents do. And Pat, in particular, could take pride in the ultimate gift of love that she had given to her daughter. Finally, David, Jose, and Paula will grow up with the knowledge that they have a special grandmother unlike almost any other grandmother in the world.

WHEN THINGS GO WRONG:
MARY BETH AND BILL

Among all reproductive technologies practiced now, and in the future, surrogacy will always hold a unique position at the heart of a potential ethical dilemma that can pit parent against parent. To appreciate the uniqueness of this dilemma, we must first explore the emotional attachment associated with each component of the reproductive process.

Of the three biological ingredients that go into the creation of a baby, the sperm component is the easiest to come by. A fertile man can deposit 100 million viable sperm into a cup in a few minutes without any third-party assistance. With today's technology, the contents of that single cup could be used to repopulate the entire country of France.

Eggs are not so easily obtained. But if a woman is willing to subject herself to two weeks of hormonal treatments and a simple medical procedure that can be carried out in a doctor's office, it is possible to recover a dozen or two of her gametes as well. The increase in time commitment and discomfort experienced by an egg donor is typically balanced by a much larger reimbursement for services.

Men and women who contribute sperm or eggs alone to collaborative reproduction are similar in an important way. Both can be eliminated from the process before a single embryo is conceived, and even before it is known which of the many gametes they provided (if any) might actually be used in the formation of a child. As a consequence, sperm and egg donors need never have a chance to form an emotional bond with a potential descendant, except perhaps in their imagination.

The third biological ingredient in reproduction—the womb—differs substantially from the other two in the commitment it entails from the woman. A surrogate mother may be forced to make physical and emotional sacrifices that far exceed any made by either gamete donor. And unlike sperm and egg donors, a surrogate mother may try to back out of a collaborative reproduction arrangement at the very last minute and keep the child that is born. For these reasons, surrogacy is often seen by potential parents as a high-risk approach to overcoming an infertility problem.

What might cause a woman to renege on a surrogacy arrangement after a birth has taken place? It's the intimate contact that exists between the surrogate mother and her maturing fetus. The fetus "never leaves her side" throughout her pregnancy. By the end of the fifth month, she can feel it move as she eats, sleeps, and goes about her daily activities. During the last two months of her pregnancy, she can feel it respond to external stimuli like loud sounds or music. At the end of her pregnancy, she can watch as this fetus emerges from her body as a baby. Although she agreed to give away a "virtual baby" before it was even conceived, natural instincts—accumulated through millions of years of evolution—may be tell-ing her now to renege on her promise and keep her very real baby for herself. (It is important to point out that even during the early years of commercial surrogacy, less than 1 percent of surrogate birth mothers reacted in this way.)

During the same nine-month period, there is another couple anxiously awaiting the birth of *their* baby. The man may be the *father* of this baby according to every meaning of the word. The woman will be only a prospective mother in the case of traditional surrogacy, but she will be the gene-mom, as well, in the case of gestational surrogacy. Although neither will feel the fetus within their bodies, they will almost certainly feel it within their minds. In a very real sense, gestational surrogacy can provide a prospective mother with an experience parallel to that felt by billions of prospective fathers, as they await the births of their children during the course of normal reproduction.

What happens when the surrogate mother and the contracting couple both want to keep their baby? How does one resolve a conflict between what may be equally valid, but incompatible expressions of essential human desires? The nation was forced to consider this very ethical dilemma when Mary Beth Whitehead decided that she wanted to keep the child that she bore under contract to William (Bill) and Elizabeth (Betsy) Stern.

Bill and Betsy were forty-one-year-old professionals who lived in the upper middle class New Jersey suburb of Tenafly. They had been married for twelve years when they decided to use surrogacy as a means to have a child. Bill was a biochemist and Betsy a pediatrician. Betsy had no reason to believe that she was infertile, but she had a mild case of multiple sclerosis and feared that a pregnancy could exacerbate her physical condition. In 1985, the Sterns contacted the surrogacy broker Noel Keane who brought them together with Mary Beth Whitehead, an unemployed twenty-nine-year-old mother of two (ages ten and twelve) married to a sanitation worker and living in a working-class neighborhood of Brick Township, a short drive away from Tenafly. Keane drafted a surrogacy contract, signed by the Whiteheads and the Sterns, in which Mary Beth agreed to achieve pregnancy through artificial insemination with Bill's sperm, and then relinquish the baby to the Sterns after it was born. The Sterns paid a broker fee of $7,500 at the outset, and agreed to reimburse Mary Beth for all costs associated with the pregnancy and pay her $10,000 after the baby had been transferred to their custody.

"Baby M," as she is referred to by the court, was born on March 27, 1986. At once, Mary Beth knew she had a made a mistake. "Seeing her, holding her, she was my child," she said. "I signed on an egg. I didn't sign on a baby girl, a clone of my other little girl." With these thoughts in mind, she put the name Sara Elizabeth Whitehead on the birth certificate.

Nevertheless, after a three-day stay in the hospital with her birth mother, the baby was handed over to the Sterns according to the contract. The Sterns named the little girl Melissa. But less than twenty-four hours later, Mary Beth appeared at the front door of the Sterns' Tenafly house and pleaded with them to allow her to bring the baby back to her Brick Township home for just a week. The Sterns agreed, hoping that a short period with her baby would allow Mary Beth to come to terms with her original decision and relinquish the child peacefully. But a peaceful resolution was not to be had.

Mary Beth refused to accept the $10,000 surrogacy fee and for five weeks kept Sara at her home, even as the Sterns pleaded for the child's return. Finally, the Sterns obtained a secret court order granting them custody of Melissa. Together with court marshals, the Sterns stormed into the Whiteheads' home to rescue their baby. But as the Sterns waited in the living room, Sara was passed out through a ground floor bedroom window and whisked away down the New Jersey Turnpike. When the marshals realized that the baby was gone, there was nothing more they could do at the Whitehead house, and the Sterns were forced to return home in horror and anguish. Soon thereafter, Mary Beth disappeared from her New Jersey home as well.

A private detective was hired by the Sterns to search for the missing baby. For three-and-a-half months, Bill and Betsy waited in Tenafly, not knowing whether they would ever see Melissa again. Finally, the detective discovered Mary Beth and her baby hiding with Mary Beth's mother and other family members in a small home in Florida. The FBI was notified and surrounded the house as the detective went inside and found Sara nursing at her mother's breast. As the law watched, he pulled the baby from Mary Beth, whisked her out of the house and back north to Bill and Betsy Stern. It was now the Whitehead family's turn to feel horror and anguish.

As the Sterns maintained custody of Melissa, the case proceeded to a highly publicized six-week trial. One year after the baby's birth, the judge, Harvey Sorkow, announced his decision. He accepted the surrogacy contract as valid and enforceable; granted permanent custody of the child to Bill Stern; terminated all of Mary Beth's parental rights and ties to the child; and made Betsy Stern the legal mother.

Mary Beth was appalled by the ruling, and she appealed it before the New Jersey Supreme Court. On February 3, 1988, the high court made its decision "In the matter of Baby M." Although mostly a victory for the Sterns, it was at the same time a defeat for the future practice of commercial surrogacy in the state of New Jersey. What the court did was to overturn the trial court decision and declare all surrogacy contracts, like the one entered into by the Sterns and Whiteheads, illegal and unenforceable. The practical consequence was that if a woman did decide to enter into a now-illegal surrogacy contract, she was under no obligation to relinquish the child that she had conceived for the contracting couple or individual.

But in the particular "matter of Baby M," the court decided to grant custody to Bill and Betsy, primarily because the baby had already lived with them for almost two years and it was deemed in the child's best interest for her to stay with the man and woman she called daddy and mommy. At the same time, Mary Beth Whitehead was granted weekly visitation rights and regained her status as the legal mother. While the outcome was less than ideal for all concerned, there was no appeal to the U.S. Supreme Court; everyone felt that Melissa/Sara had already experienced enough turmoil to last two lifetimes.

IN THE WAKE OF BABY M

There is no simple ethical solution, no single right answer, to the dilemma that arises when a surrogate mother wants to keep her baby. As such, it is fascinating to look at the political alignments that have occurred with conservatives and liberals on both sides of this issue. Prominent feminists like Betty Friedan and Gloria Steinem, who feel that surrogacy degrades women, have allied themselves with strange bedfellows—like the social reactionary Senator Henry Hyde and the Catholic church—to oppose the validity of surrogacy contracts and argue in favor of the birth mother keeping the baby. In contrast, a small number of feminist scholars like Lori Andrews—who oppose governmental restriction on women's procreative choices—are allied with traditional conservatives who believe that fully informed individuals should live up to the terms of the contracts they've signed.

Others—including individuals and state courts—take a more nuanced view of the biology and make their decision accordingly. To those in the middle, the double biological contribution—of genes and a womb—made by traditional surrogates like Mary Beth Whitehead trumps the single contribution made by a contracting biological father like Bill Stern. In cases of gestational surrogacy, the balance tips the other way—with the contracting couple providing both the maternal and paternal genetic components, and the surrogate mother providing only the womb. Many assume (falsely) that there can only be a single "real biological mother," and they award this distinction to the contracting genetic mother (who only contributed an egg). With this analysis, the baby that results from gestational surrogacy is naturally given to the contracting couple.

Although it was clear that ethical dilemmas of the type raised by Baby M would only occur in a small fraction of surrogacy arrangements, the televised plight of Mary Beth Whitehead had an enormous legal fallout. In addition to New Jersey, a number of states, including Arizona, Kentucky, Indiana, New Mexico, North Dakota, Louisiana, Nebraska, Utah, and Washington have ruled that surrogacy contracts are void and unenforceable. What this means, in practice, is that a surrogate mother living in any one of these states has the right to change her mind after a birth and retain custody of the baby. The law in Arizona went so far as to establish the surrogate mother and her husband (if she is married) as the legal mother and father of a child, even in the case of gestational surrogacy where sperm and egg were both derived from a contracting couple. (This law was recently struck down in the Arizona courts.)

Some states go even further. The legislatures of Michigan, Virginia, New Hampshire, and New York have all ruled the practice of commercial surrogacy a criminal act with the imposition of penalties on those who take part. In Michigan, the penalties are severe. Surrogate mothers themselves can be charged with a misdemeanor that carries a fine of up to $10,000 and one year in prison. And individuals who arrange a surrogacy agreement can be charged with a felony, which carries a fine of up to $50,000 and five years in prison. Policies outside the United States are similar. The United Kingdom, France, Germany, and Australia all criminalize commercial surrogacy to one degree or another.

Mary Beth Whitehead is the symbol of all that can go wrong with a surrogacy arrangement and the rallying point for those who would ban this component of reprogenetics. But what if surrogacy could be practiced in a way that eliminated the possibility of any future Mary Beths? What if it became possible to guarantee that every contracted surrogate mother would freely, and without regret, relinquish the baby she conceived to a contracting couple or individual? Would this elimination of "the real problem with surrogacy" make a difference, if not in the law, at least in the real world?

THE REALITY

Surrogacy has not been in the news much lately. There is the occasional case with a peculiar twist that shows up one day and is forgotten the next, but in general, the surrogacy front seems very quiet. Perhaps, you might think, the law has taken its toll and successfully removed this reprogenetic practice from the American landscape. Perhaps it has been run underground to be practiced only on occasion and always away from the prying eyes of the reproductive police.

Think again.

The American surrogacy industry is booming and it's in broad daylight. If you cruise onto the World Wide Web and go to the address http://www.surrogacy.com, you'll find The American Surrogacy Center, Inc., which lists surrogacy brokers and full-service agencies on a state-by-state basis. In August 1996, there were seven listings in California alone. The on-line site also provides information, articles, classified advertisements, and detailed descriptions of all aspects of the surrogacy process. Surf over to http://www.opts.com, and you'll find The Organization of Parents Through Surrogacy, a "national, nonprofit, all-volunteer organization whose purpose is mutual support, networking, and dissemination of information regarding surrogacy." And for detailed information on individual surrogacy agencies go to http://www.surroparenting.com for the Center for Surrogate Parenting & Egg Donation in Beverly Hills, California, or http://www.babies-by-levin.com for Surrogate Parenting Associates in Louisville, Kentucky, or http://www.surrogacyagency.com for Surrogate Parenting Center of Texas in Austin, or http://www.surrogatemothers.com/ for Surrogate Mothers in Monrovia, Indiana.

If commercial surrogacy is being practiced to a greater extent than ever before, why isn't it in the news anymore? First, it seems that everything that could be said about surrogacy has been said ad nauseam. Second, surrogacy facilitators have figured out how to avoid the well-publicized fiascos of the past through judicious screening of potential surrogate mothers.

But how can surrogacy be practiced in broad daylight across the United States when so many states have made it illegal? It's all because of the American Constitution. The Constitution clearly hands over to individual states the right to decide all laws that govern family matters such as marriage, divorce, adoption, and birth rights. This means that if you want to get married or divorced quickly, you can find a state that will satisfy your wishes. And if you want to use a surrogate mother to give birth to your child, the same holds true.

The practical effect of the governance of surrogacy by state law is that the only laws that count are the ones in the most lenient states. There's no reason to risk hefty fines and jail time in Michigan when you can drive across the border to Ohio where a gestational surrogate will be ordered by the court to live up to the terms of the contract that she signed before she got pregnant. And the cost of several cross-country plane tickets between New Jersey and California is minimal compared to the $45,000 that you may spend no matter where the surrogate lives.

Perhaps surprisingly, the most surrogacy-friendly state in the country today—as far as contracting parents are concerned—is Arkansas. In cases of both traditional and gestational surrogacy, Arkansas recognizes the contracting couple or individual as the legal parents or parent whose name(s) are placed on the birth certificate. If a dispute were to arise, the contract would control the outcome, which means that the surrogate mother would have to relinquish custody of the baby.

Two other surrogacy-friendly states are California and Ohio, where contracting parents are deemed the legal parents in cases of gestational surrogacy. In these states, however, a surrogate mother in a traditional arrangement might be able to claim custody over the child if a dispute were to arise. But imagine what would happen if it became possible for surrogacy agencies to guarantee that the birth mother would freely relinquish the baby at the time of birth. The number of states where commercial surrogacy could be practiced out in the open would increase dramatically to include all those where it wasn't expressly forbidden. This would likely include up to half the states in the union.

For all practical purposes, we are already there. The Beverly Hills

Center for Surrogate Parenting & Egg Donation has shepherded 456 surrogate births as of April 1996, all without incident, and in Kentucky, Surrogate Parenting Associates has shepherded more than 500 surrogate births, again all without incident.

These and other surrogacy centers have reached the goal of almost guaranteed success by carefully selecting surrogate mothers based on an extensive series of criteria that are applied to each potential surrogate through interviews and psychological screening. The Center for Surrogate Parenting & Egg Donation says that it only accepts one prospective surrogate mother out of twenty that apply, and it bases its selection on a strict screening process that can take three to five months. The current status of commercial surrogacy is aptly summarized by Steven C. Litz of The American Surrogacy Center:

No surrogate should be allowed to participate in a program without thorough psychological screening. While some programs are notorious for their lack of screening, they are also the ones who have had disasters occur. When surrogacy is conducted responsibly, it is universally successful. Of the thousands of children born to surrogates, there has yet to be a single case where a surrogate who was adequately screened before conception changed her mind and tried to keep the child.

Can we use the evolving practice of surrogacy in America as a harbinger of the future use of more advanced reprogenetic technologies? It is my contention that we can, and the implications are stunning. What the brief history of surrogacy tells us is that Americans will not be hindered by ethical uncertainty, state-specific injunctions, or high costs in their drive to gain access to any technology that they feel will help them achieve their reproductive goals.

13

Buying and Selling Sperm and Eggs

DONOR INSEMINATION

When the missing ingredient in a couple's desire to have a child is the father's genetic material or when a single woman desires to have a child without the participation of a man, the solution to babymaking is artificial insemination by donor, now more commonly referred to as donor insemination or DI.

Artificial insemination has a long history as a reprogenetic technology, going back more than two hundred years to a time before it was even realized that a single sperm cell was required to initiate a pregnancy. The Italian priest and physiologist Lazzaro Spallanzani was the first to achieve a pregnancy with this technique in experiments performed on dogs in 1782. And in the 1790s, the Scottish physician John Hunter facilitated the first successful artificial insemination of a woman with her husband's sperm. In this particular case, the man was infertile solely because of a congenital malformation of the penis. Hunter didn't do the procedure himself; instead, he gave the husband a syringe to be filled with ejaculate for injection into his wife's vagina.

Additional cases of artificial insemination of women with their husband's sperm (as a medical treatment for certain fertility problems) were reported in various countries of Europe and in the United States throughout the nineteenth century. And in 1884, at Jefferson Medical School in Philadelphia, the first successful insemination of a married woman with sperm from an anonymous donor was performed by the physician William Pencoast. The donor was a medical student described as "the handsomest student in his class."

The use of donor insemination grew steadily, and quietly, during the years before World War II. But only a fraction of the couples who might have taken advantage of the technique actually did, for most husbands resisted the notion of having and raising a child that was not genetically their own. In many societies and cultures, a marriage is not perceived to be truly consummated until a couple reproduce children with their *own* genetic material. In this light, a DI child would forever pose a threat to a man's ego as living proof that he was unable to perform "nature's primary task."

Another reason for the slow acceptance of DI was the shocking nature of this first of the many reprogenetic technologies to be developed and used during the twentieth century. Some legislatures and courts in the United States, Canada, and England equated DI with adultery and DI children were labelled illegitimate, even if the husband had consented to the procedure. The negative legal response to DI that held sway into the 1960s was clearly a mirror of prevailing public opinion in an era when "good girls" waited until marriage before having sex, and divorced women were considered "used goods." The thought of having a strange man's semen in their vagina was probably enough to stop many women from considering DI.

In addition to these legal and emotional problems was a technical one imposed by the difficulty of coordinating the schedules of the sperm donor and the recipient couple. The date on which the procedure could be performed was determined entirely by the woman's natural ovulatory cycle. And on this date, the sperm donor had to appear *in person* in the physician's office, or close by, and masturbate on demand. As described by fertility experts R. Snowden and G. D. Mitchell:

The donor may not be feeling particularly well or it may be otherwise
inconvenient to supply the semen at a given time. The process of mas-
turbation to order during a working day is not an activity that most
men would consider satisfactory. . . . The donor is also asked to abstain
from sexual activity for a period of time before the semen is collected
and the inconvenience of this may be considerable. . . . The donation
of semen is not as straightforward or as easy to arrange as it might
appear.

With the first report of a technique for the successful cryopreservation
of human sperm in 1953, scheduling problems for the donor could be
eliminated. Suddenly, it became possible to store many different sperm
samples ahead of time with donations obtained on different days from
different men. The freezing technology provided physicians with the ability
to select the donor with physical characteristics that most closely matched
those of the husband, so that families could more readily hide the origin
of their DI children from the world. Furthermore, sperm donor and recipient never had to be in the same place at the same time, which could serve
to eliminate fears of an accidental meeting and ensure donor anonymity.

But although sperm freezing eliminated technical problems associated
with DI, and its use gradually increased, it was still practiced in the shadow
of the law and public opinion for the same reasons that prevailed before
1953. Then, in the 1970s, the use of DI by infertile couples skyrocketed
as it became generally accepted across American society.

A new code of sexual morality, an increased openness in discussing
male infertility, and a reduction in the pool of babies put up for adoption
played a role in the change of public opinion. Again, a critical factor was
simply the passage of time required for a practice viewed initially as alien
and unnatural to become familiar and nonthreatening.

In 1974, the American Medical Association made an ill-advised attempt
to take control over this burgeoning new reprogenetic technique by pro-
posing that only licensed physicians be allowed to practice DI. Unfortunately, most male physicians of the day tried to impose their own moral
code on which women could be inseminated, and this almost always excluded unmarried women and lesbians.

The AMA effort to corner the market was bound to fail anyway since
the essential ingredients required for DI—a plastic container to hold the
ejaculate and a turkey baster to deposit it near the cervix—can be pur-

chased at a supermarket. And beginning in 1978, a series of women-friendly health centers established donor insemination programs that allowed single heterosexual and gay women to obtain anonymous sperm donations as well.

A comprehensive 1987 U.S. government survey of artificial insemination showed how widespread and accepted the practice had become. Over the span of a single twelve-month period between 1986 and 1987, an estimated 8,000 physicians offered donor insemination services to 77,000 women, which resulted in the birth of approximately 30,000 DI babies. Add to this number all of the DI pregnancies achieved each year without physician assistance, and extrapolate out to the mid-1990s when the birth rate is even higher, DI is more fully accepted by society, and single and gay women are becoming parents in unprecedented numbers, and you get an idea of the enormous impact that this simple reproductive technology has had on society. It is impossible to know the total number of people alive today who were conceived through DI, but an estimate of 1 million does not seem unreasonable.

EGG DONATION

So far we have considered methods that can be used by couples missing either the sperm ingredient or the womb ingredient from the reproductive equation. The final ingredient in this equation is the egg. When the male member of a couple is fertile and the female member is unable to produce viable eggs, but is otherwise healthy, there are two potential ways to have a child who is genetically related to the social father. The couple can engage the services of a surrogate mother, as we discussed previously, or they can obtain donated eggs for fertilization, introduction, and hopefully implantation, into the uterus of the woman who desires the child.

In contrast to artificial insemination, which has been practiced for more than two hundred years, and surrogacy, which has been practiced for at least three thousand, egg donation only became possible with the successful application of human IVF beginning in 1978. The first pregnancy to be initiated with a donated egg was reported at Monash University in Melbourne, Australia, in November 1983. Since that time, the practice has spread around the world to many of the same clinics that practice standard IVF.

Egg donation has several advantages over surrogacy. First, with the removal of the egg donor from the scene before fertilization even occurs, the possibility of an attachment forming between donor and child can be eliminated. Second, the infertile woman can become, in at least one sense, the biological mother of the child that is born. Finally, the number of women willing to donate eggs is much larger than the number of women willing to become surrogate mothers. So while couples may have to wait months or years to find a suitable surrogate, the Center for Surrogate Parenting & Egg Donation reported in December 1995 that they had a list of 150 women who were *waiting* for recipient couples to choose them as egg donors.

What this fertility center and others have begun to do, in fact, is to treat egg donors in almost the same way that sperm banks treat sperm donors. Egg donors are recruited by advertising or "word of mouth" and are interviewed by specially trained psychologists and social workers at the fertility center. Those who meet established criteria are put on a list that is presented to prospective recipient couples. In the one main difference from sperm donation, egg donation typically does not ensue until *after* a particular donor has been chosen by a recipient couple. This allows freshly donated eggs to be fertilized immediately with the prospective father's sperm.

OTHER COMBINATIONS AND POSSIBILITIES

Donor sperm, donor eggs, and donor wombs can be, and have been, combined in every possible way to provide couples and individuals with the opportunity to overcome different combinations of infertility. When couples are unable to produce either sperm or eggs, but the woman has a healthy uterus, they can resort to the use of either a donor embryo or two donor gametes. Donor embryos are often available from other infertile couples who have succeeded at IVF and have leftover frozen embryos that they don't plan to use. But a couple can gain more control over the process by choosing a specific egg donor and a specific sperm donor (who will never actually meet each other) to generate fertilized eggs by IVF that are introduced into the womb of the prospective social mother.

Since the child that emerges from a double donor arrangement has no genetic connection to either social parent and, in this sense, is no different from one who has been adopted, you may ask, "Why bother?" Because by using reprogenetic technology, a woman is able to establish—with a child—a biological connection that she would not have had in the case of adoption, and because it is much more difficult today to find healthy newborn children to adopt than it is to find sperm and egg donors. Thus, an embryo or double gamete donation is probably the easiest route that such couples can take to become parents.

In another combination of fertility problems, the female member of a couple may be unable to carry a fetus to term, but will be able to produce eggs, while her male partner will be unable to produce sperm. The simplest solution in this case, other than adoption, is to engage a gestational surrogate mother who receives embryos formed in vitro with eggs provided by the social mother-to-be and sperm provided by a donor. This combination of reprogenetic technologies turns the traditional surrogacy arrangement on its head, since the baby that emerges is genetically related to the contracting mother but not the contracting father.

When a woman is unable to produce eggs and is also unable to carry a fetus to term (owing to a radical hysterectomy, for example), she will again need the services of a surrogate. If her male partner is fertile, the simplest solution would be to contract for a traditional surrogacy arrangement in which the surrogate is artificially inseminated with the prospective father's sperm, as in the case of Baby M. However, legal experts on surrogacy like Thomas M. Pinkerton, writing for The American Surrogacy Center, have suggested the use of donated eggs from a third woman in such cases to reduce the connection between the surrogate mother and the baby she brings to birth. This arrangement is referred to as "ovum-donor gestational surrogacy." In the unlikely event that a surrogate in such an arrangement changes her mind, the courts, in California at least, would presumably grant custody of the child to the contracting couple since the surrogate would have no genetic tie to the baby. In a traditional surrogacy arrangement in California, the surrogate mother would be given preference instead.

Finally, one can imagine a scenario in which the three biological ingredients come from three different donors, when both members of a couple are unable to produce sperm or eggs, and the prospective social mother is unable to carry a fetus to term. They contract with a surrogate mother to be implanted with embryos formed in vitro with donated sperm and eggs. The child who is born will have to contend with having a social mother and social father and also a biological father, a genetic mother, and a birth mother—five different primary parents in all!

I have no doubt that such scenarios have occurred already and will occur more frequently in the future. Why? First, there is the legal advantage suggested by Pinkerton in severing the genetic connection between the surrogate mother and the fetus she carries. But, in addition, there is another perceived advantage to ovum-donor gestational surrogacy that most people think is better left unsaid. This is the notion that women willing to become surrogate mothers are likely to be missing some important attributes that infertile couples would prefer to see in the genetic mother of their child.

A good surrogate mother is one who already has children of her own and is willing to put up with the discomforts of pregnancy in order to benefit substantially from a fee of $15,000 or less. Typical surrogate mothers will not have college degrees or professional careers. Although a woman in good health with just a high school education may make an ideal surrogate mother, she will seem less than ideal as a genetic mother to many highly educated dual-career couples in need of an egg.

And who would an ideal gene-mom be? Perhaps an undergraduate major in physics at Princeton University. The following advertisement ran in Princeton's student newspaper during April 1995. Similar ads are commonly listed in student newspapers at every Ivy League university: "Loving, infertile couple (Yale '80 grad and husband) wanting to start a family needs a healthy, light-haired, Caucasian woman (ages 21–32) willing to be an egg donor. Reimbursed $2,000 plus expenses for time and effort. Comprehensive physical at leading NYC hospital included. Please call (212) . . ."

GAMETE DONATION AND EUGENICS?

As the practice of donor insemination grew slowly from the 1930s through the 1970s, physicians followed along the trail blazed by William Pencoast in 1884. With few exceptions, they were the ones who chose the donor, and more often than not, the donor was a medical student, or another doctor. In addition, many practitioners selected only those who appeared to possess superior intelligence, a pleasing personality, and "good looks," while nearly all rejected anyone with personality or health problems of any kind. In more recent times, physicians have subjected prospective donors to family histories as well as complete physical exams to further weed out those who might possibly carry disease genes or who have more subtle physical or mental health problems.

Although each of the thousands of physicians who have performed DI over the years may have acted solely in regard to his or her own conscience, in the aggregate, they have caused more than a million human beings to be born with a combined genetic heritage that is clearly skewed from the average for the population as a whole. While the contribution of genes to innate intellectual abilities in any individual may be debated, all contemporary scientists agree that some contribution does exist, and there is no longer any question that genes play a fundamental role in physiological and mental diseases—such as obesity, schizophrenia, alcoholism, and manic depression—as well as general physical characteristics, both positive and negative. Thus, what DI practitioners have done through their methods of donor selection is to shift the distribution of genes present in the million-strong population of DI-conceived Americans away from the distribution present in the population at large.

Plain and simple, the practice of donor insemination in America has produced a *eugenic* outcome, according to the original definition of this word. Just as Francis Galton, the founding father of eugenics, proposed in the late 1800s, "selective breeding" among humans has been used to alter the gene pool of a small portion of the population.

Without a doubt, governmental attempts to impose eugenic policies have caused harm to individuals as well as to whole societies through restrictions on immigration and reproductive freedom and, in the most extreme cases, genocide. But DI eugenics does not restrict individual reproductive possibilities, it enhances them. The children who are born are not harmed in any way. In fact, from a statistical point of view, they are less likely to develop the disease traits that have been selected against in donors. The parents are not harmed either, so long as they don't hold unrealistic expectations of what their children can achieve (which is a problem for many parents of traditionally conceived children). Finally, it is difficult to see any negative effect that donor selection might have on society as a whole.

If there were any problem with the traditional practice of physician-assisted DI, it was that the doctor (almost always a male) typically assumed that he was the one who could make the best choice for the couple in need. But now DI is becoming a consumer market with more and more prospective parents deciding for themselves on the choice of a donor from catalogs provided by numerous American sperm banks that are competing for their business.

The trend toward a consumer market began with the establishment, in 1979, of a California sperm bank called The Repository for Germinal Choice, founded by a retired seventy-five-year-old optometrist named Robert Graham, who had made a fortune from his invention of shatterproof eye glasses. Graham, who funded the nonprofit sperm bank with his own money, intended originally to accept sperm donations only from Nobel Prize winners and to provide samples only to well-adjusted married women with IQs over 140.

Graham's intentions were clearly eugenic. As echoed by one of his donors, the controversial Nobel laureate William Shockley, "The principles of this may not be popular, but they are sound. We're trying to take advantage of the possibilities of genetics. We are hoping for a few more creative, intelligent people who otherwise might not be born."

The media response to Graham's project was one of horror and amusement. The "Nobel Sperm Bank," as it was routinely called, was the butt of many jokes as well as critical editorials in newspapers across the country. The editors of the *New York Times* remarked sarcastically, "If intellectual qualities were inheritable in any simple fashion, those who conceive with the help of the Nobel sperm could count on offspring endowed with a great deal of vanity and a plain dearth of sense. Chances are, however, they will get themselves just children."

Other critics were not so kind. Dr. Kenneth Dumars, head of the division of clinical genetics at the University of California Irvine, declared, "This is a gimmick, an unrealistic hope for families. To hold out the idea that Nobel sperm will help society is sheer bull."

By 1984, Graham was forced to abandon his original plans. He had been unable to convince more than a handful of Nobel Laureates to donate their sperm, and the few samples obtained were of poor quality owing to the advanced age of the donors. Surprising to Graham, though, was his discovery that most women were not interested in being inseminated by Nobel sperm. "Nobelists are generally so old that all the female recipients were turned off," he said. "Even on paper, women are drawn to younger men."

So he solicited young California research scientists whom he called "potential Nobelists" as well as others who excelled physically like a Gold Medal–winning Olympic athlete. He also decided to relax his requirements for recipient women, allowing almost anyone who was married, under thirty-eight, healthy, and able to provide a child with a decent standard of living to participate.

By this point, Graham's approach to Donor Insemination was different only in degree from the selective protocols used routinely by most other DI providers who also chose men of high intelligence and good physical characteristics. The real difference in the early 1980s was that Graham allowed couples to choose their donor, while most other DI providers did not.

By the mid-1990s, other selective sperm banks catering to the consumer sprang up around the country. Some, like The Fertility Research Foundation in Manhattan, also stock samples from scientists and Olympic athletes. Others, like Cryobank, with branches in Palo Alto and Boston, specialize in sperm samples from top-ranked college students at places like Stanford, MIT, and Harvard. Prospective customers are provided with a catalog describing pertinent characteristics of each donor, including a detailed physical description; medical history; special talents in music, art, or athletics; education; and SAT scores. For an additional fee, a "donor matching counselor" can be enlisted to provide a more sophisticated match between the desires of the recipient couple and a chosen donor.

The Repository for Germinal Choice (which had facilitated two hundred births by the end of 1991) and other sperm banks will not change society in the grandiose way that Graham imagined. Instead, they will simply provide infertile couples with the opportunity to select donors having characteristics that *they,* as prospective parents, find appealing. The chance that those same characteristics will end up in the children that result is not at all guaranteed.

If presented with the choice, the vast majority of prospective parents would rather give their children their own genes and not those of a gamete donor, irrespective of who that donor might be. But when a person is unable to pass his or her genes on to his or her child, then gamete donation becomes an alternative. And many people in Western societies today would prefer to raise "as their own" a child with a stranger's genes rather than have no child at all.

Once the decision has been made to accept gamete donation, it seems only reasonable to assume that parents will want to select the best donor possible. What does this mean? Consider the comments made in 1994 by several satisfied parents who had conceived children with sperm samples obtained from Graham's sperm bank:

> *"First of all, we wanted a healthy baby. But we also wanted a special baby, someone who would do well, someone who would succeed. . . . Doesn't every parent want that? Doesn't everybody want their baby to be smarter than the others?"* (Sandy Fuller)

> *"At the end of the day, I only wanted a healthy, happy baby—but why not have a child born with an advantage in this increasingly difficult world?"* (Afton Blake)

> *"Is drawing the wild card always so wonderful?"* (David Ramm)

Even the bioethicist Arthur Caplan, who originally criticized Graham's enterprise as "morally pernicious," was forced to agree: "We mold and shape our children according to environmental factors. We give them piano lessons and every other type of lesson imaginable. I'm not sure there is anything wrong with using genetics . . . as long as it is not hurting anyone or he (Graham) is not imposing his ideas of perfection on anybody."

14

Confused Heritage

YOUR "OWN" CHILD

From the time our ancestors first understood the connection between sex and reproduction, a mother understood her own child to be the one she gave birth to, and a father's own child was the one conceived with semen that he deposited into a woman's vagina. It was on the basis of this clear distinction that the desire to have "one's own" children became programmed into our genes—over the course of evolution—as a natural instinct.

The distinction made between "one's own child" and "someone else's child" throughout history was much greater than many now realize. Adoption of unrelated children was extremely rare until early in the twentieth century. Children orphaned without relatives may have been cared for by foster parents in earlier times, but such parents invariably distinguished between their "own children" and the children of others.

With the use of reprogenetic technologies, the meaning of "one's own child" becomes blurred. For IVF makes it possible for one woman to be the birth mother to a child conceived with another woman's egg. Which of these women has the right to consider the child her own?

What most educated citizens of the Western world in the late twentieth century would say is that the child "belongs to" the woman whose egg was used in its conception. Infused, as we are, with a sophisticated understanding of biology, we know that all of the child's inherited characteristics are carried in the egg and sperm; none are contributed by the birth-mother's blood or body. Furthermore, we know that these characteristics are programmed by the *genes* present within the fertilized egg. We speak confidently of a genetic mother who can rightfully call a child born with her genes her "own child," no matter where its development took place. We place an intellectual veil over our primitive instincts in order to accept the birth of "our own child" through the birth canal of another woman.

Are genes really the only determining factor in considering who a baby "belongs to"? Or is it really not that straightforward?

TWIN SISTERS CAN PROVIDE THE ULTIMATE GIFT TO EACH OTHER

Florence and Gail are identical twin sisters. Florence got married to Frank, and Gail got married to Gary. Unfortunately, before she even met Frank, Florence developed ovarian cysts, which necessitated the surgical removal of both of her ovaries. Florence and Frank now want to have children but Florence is unable to produce eggs. To help her sister out, Gail agrees to donate her eggs to Florence. Gail's eggs are fertilized *in vitro* with Frank's sperm and introduced into Florence's uterus. Nine months later, Florence gives birth to a baby girl she names Fiora.

Who is Fiora's genetic mother? It's Gail, of course, you would say, since she contributed the egg that developed into Fiora. But, in fact, if Fiora and her birth mother Florence were subjected to DNA fingerprint testing, the results would be quite definitive—they would show, without question, that Florence herself was Fiora's gene-mom. What's going on here?

The confusion is caused by Florence and Gail being identical twins. As a consequence, they have exactly the same genes. Every egg that Gail produces carries half her genes. But any one-half portion of Gail's genes is equivalent to a one-half portion of Florence's genes. Thus, the eggs produced by Gail could all have been produced by Florence.

Another way of looking at this is from the point of view of the single fertilized egg that developed into both Gail and Florence. This single cell underwent about a hundred divisions and then a small number of its descendant cells reduced their genetic material by half to become eggs. Some of these eggs ended up, by chance, in Gail's ovaries while others ended up, by chance, in Florence's ovaries (which were surgically removed).

In strictly genetic terms, Gail and Florence must both be considered Fiora's genetic mother. But this conclusion is rather unsettling, because it means that by DNA fingerprint analysis, the children of *all* identical twins will be found to have two genetic mothers or two genetic fathers—their social parent and their aunt or uncle. It also means that all first cousins related through identical twin parents will appear to be half-brothers or half-sisters.

The children of an identical twin mother don't normally think in this way for a very simple reason. Their social mother is also their birth-mom as well as their gene-mom, while their aunt is connected only by genes. But what about Florence and Fiora? Florence is a gene-mom, she is the birth-mom, and she intends to be the social mom of Fiora. Does this combination trump Gail's contribution of an egg that Florence could have produced herself if she had ovaries? The only unique contribution made by Gail is that of storing the egg for some twenty-five years before graciously handing it over for use by her sister.

Let's consider another scenario that is similar but goes beyond semantics to a question of medical approach. This time the identical twin sisters are Amy and Jane. Amy is married to Andrew and Jane is married to Jay. Amy has a uterine infection that forces her to have a hysterectomy, but her ovaries remain intact and functional. Amy and Andrew want to have "their own" children, and Jane has agreed to act as a gestational surrogate mother. Amy plans to have her eggs recovered for fertilization in vitro with her husband Andrew's sperm. The fertilized eggs will then be introduced into Jane's uterus for implantation. Jane will carry the fetus to term and then give the baby over to Amy and Andrew so that they can raise their "own child."

What we learned from the previous scenario of Florence and Gail is that identical twin sisters can both be considered genetic mothers of any child conceived from eggs produced by either woman. This means that a child conceived by in vitro fertilization with Amy's egg and Andrew's sperm would have the same genetic heritage as one conceived through the fertilization of Jane's egg by Andrew's sperm, which could be accomplished by artificial insemination.

What does Amy do? Artificial insemination is cheaper and much less intrusive than IVF for both women. The child born in either case will have the same birth-mom and the same pair of gene-moms. So what difference does it make?

Amy may try to argue that although she and her sister share the same genes, she wants to use her egg so that her child receives the particular DNA molecules that she produced in her own body. Surprisingly to Amy, this argument doesn't work because, for the most part, the particular DNA molecules present in a human egg don't actually end up in the body that it develops into. Even with this knowledge, Amy may still want to contribute her *own* egg to this collaborative reproduction arrangement. Though a child conceived from Amy's egg will be indistinguishable by any imaginable test from a child conceived from her sister's egg, Amy may feel that she needs to make some *physical contribution* to her child, however ephemeral that contribution might be, and however irrational her feelings might seem to us.

In fact, rationality has nothing to do with it. It's all based on the primeval instinct programmed into Amy's genes that makes her want to have "her own child." This instinct evolved when the distinction between "one's own child" and "someone else's child" was crystal clear. And while the evolutionary purpose served by this instinct is the increased transmission of our genes to offspring, the instinct itself operates on the *physical connection between mother and child*. This is why Amy may want that physical connection instinctually even though it makes no difference to the transmission of her genes.

TWIN BROTHERS AND A CURE FOR STERILITY

Let's consider one final "twins scenario," this one actually true. The story began in 1947 on the day that Mrs. Twomey gave birth to her identical twin sons Tim and Terry. Like all pairs of identical twins, Tim and Terry looked pretty much alike and it was hard to tell them apart. But Tim and Terry were critically different in a way that was hidden from the world— Tim was born without testicles as a result of a rare developmental abnormality that had occurred while he was still within his mother's womb.

With the help of modern medicine, Tim was able to lead an outwardly normal life. At the age of 18, he began to receive weekly injections of the hormone testosterone, which allowed him to go through puberty (at a late age). And as he grew older, continued injections of the hormone provided him with the ability to engage in a normal sex life. At the age of twenty-nine, Tim married Jannie. In the meantime, Tim's brother Terry had married and become the father of three children.

At the time of their marriage, Jannie and Tim were convinced that they would never have children "of their own." For five years, Tim had been searching without success for a medical authority who could treat his fertility problem. And then shortly after his marriage, he found Dr. Sherman Silber at the Saint Luke's Hospital in St. Louis, Missouri. Dr. Silber was a urologist and skilled microsurgeon who was noted for his ability to reverse vasectomies by delicately reconnecting the severed tubes. Dr. Silber said that he might be able to cure Tim's sterility by transplanting one of Terry's testicles into Tim's scrotum. No one had ever performed such an operation before, and the obstacles to connecting both sperm and blood vessels were enormous, but Dr. Silber was convinced that he had the skills to do it successfully.

Terry and Tim both agreed to undergo the procedure, and on May 17, 1977, Dr. Silber performed the transplantation. It was a success. Within a few months, Tim achieved a normal sperm count in his ejaculate, and he no longer needed hormonal injections. On March 25, 1980, Tim and his wife Jannie—both Sacramento police officers—had a 6-pound, 14-ounce baby boy named Christopher Gene (no joke, this is really his middle name!). If genetic tests were ever performed, they would show, without a doubt, that Christopher Gene was indeed Tim's son.

How should Tim feel about this child? Should he consider Christopher "his own" son or his brother's? Would he have felt the same way if the testicular transplantation had not been possible and his child was born after his wife was artificially inseminated with his brother's sperm? Or was the production of sperm within his own scrotum necessary to set up the physical connection that allowed him to consider the child "his own"?

The facts certainly suggest that Tim would have viewed a child born by artificial insemination of his wife with Terry's sperm *differently* from the child that he gave life to himself. But why should he feel this way when "his sperm" actually came from Terry's testicle?

Again, how we think a person *should* feel rationally need not bear any resemblance to how a person *does* feel when primeval instincts prevail. Although genes drove early members of our species to desire children of their own, the kinship between parent and child was defined instinctually through the physical connections imparted by semen, gestation, and birth. Only today can we think abstractly about the genes that sit at the root of inheritance. But when intellectualization conflicts with the primeval instinct for a physical connection to one's child, we are apt to become utterly confused. There is nothing profound about this confusion. It is simply one more way in which the modern world fails to play by the rules under which we evolved.

What all three twin stories make clear is the futility of trying to come up with modern definitions for "one's own child." In the end, whether a child is one's own or not is determined simply by the way a parent feels, no matter where or how gamete differentiation or fetal development took place.

CLONING PARENTS OR NOT?

At a U.S. Senate hearing on cloning that was convened within a week after the announcement of Dolly's birth, George Annas, a lawyer and bioethicist at Boston University, warned the senators that cloning a person "would radically alter the very definition of a human being" by producing the world's first human with a single genetic parent. Was he right?

The picture that probably formed in the mind of Professor Annas was of a woman or man holding a baby who was wholly derived from a cell contributed by the adult, as in the story of Jennifer and Rachel told in chapter 9. In Professor Annas's mind, it would seem that Jennifer should be considered the genetic mother of Rachel. But what about a situation in which parents (recently infertile) decide to expand their family with a clone of a child they already have. Would the older child be the parent of the younger child, or would the two children simply be identical twins (of different ages) with the same genetic mother and father?

Professor Annas seems to be confused by the multiple types of mothers that a child can have. If Jennifer gives birth to a clone of herself named Rachel, then Jennifer is clearly Rachel's birth-mother. And if Jennifer raises Rachel herself, then she is clearly Rachel's social mother as well. In genetic terms, however, Jennifer is not Rachel's mother, she's Rachel's identical twin. This means that Rachel's genetic parents are the same as Jennifer's genetic parents. In other words, Rachel's social grandparents are also her genetic parents. And this means that Rachel and all other cloned children will always have two genetic parents, not one.

Describing the genetic relationship that clearly exists between the cloned person and the person who contributed the cell for cloning is problematic. We could say simply that they are identical twins—which they invariably are—and leave it at that. But this term fails to express the directionality of the relationship, with genetic material flowing from a person already alive toward the initiation of the life of another. To convey this special relationship, I will use the terms *genetic progenitor* to describe the person whose cell was used for cloning, and *genetic descendant* (when necessary) to describe the person who emerged from that cell. Remember, the social role played by a genetic progenitor can be that of either parent or sibling, depending on the age of the progenitor and the cirumstances under which cloning occurred.

The genetic consequences of cloning can be strange indeed. When a cloned child is raised by her adult genetic progenitor—who becomes her social mother—a generation becomes duplicated on the family tree. So when Rachel grows up and is ready to have her own children, she will have to contend with knowing that all her children will also be the genetic children of her mother Jennifer. In chapter 9, we asked whether Jennifer needed permission from her parents to produce a clone of herself in the form of Rachel. Now we can ask whether Rachel needs to obtain permission from Jennifer to have her own children through natural conception. The concern in both cases is based on the traditional assumption that a person shouldn't be forced to have a child without consent. But a logical extension of this principle would require all identical twins to request permission from their brother or sister before they had a child. Clearly, this is absurd.

Finally, there's the unusual situation that is sure to happen some day when a woman decides to clone herself after she has already had children by natural conception. The child that is born will become the genetic mother of her older brothers and sisters.

FETAL MOTHERS

It goes without saying that a *woman* is required to provide the egg and a *man* is required to provide the sperm nucleus that will come together to form a fertilized egg. Well actually, once again, our intuition betrays us. Human eggs and sperm need not come from adult women and men. In fact, they need not come from people at all!

Recall the comment made by the British-American embryologist Brigid Hogan: "I like the idea that a long time ago, in Port Elizabeth, South Africa, a young pregnant Englishwoman had inside her body not only her daughter but the egg that gave rise to her granddaughter and that the genetic recombination that contributed to me, started then."

What Hogan was alluding to is the fact that all the immature eggs, or oocytes, in a woman's ovaries are actually in place long before she is born. Six-and-a-half months before she is born, to be more precise. Amazingly, by this early stage of fetal development, the production of all of the eggs that a woman-to-be will ever have during her entire lifetime is over and done with. Some of the fetal eggs will lie in a state of suspended animation for ten to fifty years before receiving the signal to mature during a particular ovulatory cycle. Even after fifty years, an awakened oocyte still has the potential to become fertilized and develop into a new human being. Only a tiny fraction of the eggs present in the fetus will ever be called upon to go down this path. The vast majority (just like most sperm produced throughout the lifetime of a man) will just wither away.

These facts of biology tell us that every pregnant woman with a female fetus carries not only her child-to-be, but also the eggs that may ultimately develop into her grandchildren as well. The young Englishwoman in Port Elizabeth, South Africa, so long ago was Brigid Hogan's grandmother. And like all other pregnant women with a female fetus, she actually carried four ovaries filled with eggs inside her body—the two that she was born with, and the two inside her fetus.

By now, you can probably imagine a way in which this information can be put to use by modern reprogeneticists. If you can figure out a way to coax an immature oocyte extracted from a fetal ovary to mature into an egg that is ready for fertilization, the potential consequences are peculiar, to say the least.

This is just what John Eppig and his colleagues accomplished in 1995, while working at the Jackson Laboratory in Bar Harbor, Maine. With painstaking attention to the details of solutions, signals, and other required conditions, John recovered completely immature oocytes from mice and converted them into ripened eggs in a laboratory incubator. In essence, he had duplicated—in the laboratory—the maturation process that occurs naturally in the ovary when an oocyte receives the signal to proceed down the pathway of ovulation. Eppig's approach was validated when he subjected the matured oocytes to IVF, and demonstrated the development of the fertilized eggs through gestation and birth into healthy, fertile mice.

What was done in mice could be done in humans. In fact, part of the process has already been duplicated. Since 1991, reprogeneticists have been able to extract partially mature oocytes (from unstimulated ovaries), continue the maturation process in a laboratory dish, subject these dish-ovulated eggs to IVF, and end up with live-born children. It is only a matter of time before we'll be able to complete the entire maturation process with immature human oocytes (perhaps before you even read this book). And when these conditions are perfected, it will then become possible to recover ovaries from miscarried or aborted human fetuses and use the eggs within them as ingredients for the formation of new human lives. The children who are born will have genetic mothers who were never born themselves!

I admit that even the rational scientist who stands in my shoes found this concept bizarre. If we equate human life with sentience—that, of course, has yet to emerge in an early fetus—then children could be born with genetic mothers that *never existed*.

Not unexpectedly, the reactions from most bioethicists—and nonbio-ethicists—were negative. Professor George Annas exlaimed, "The idea is so grotesque as to be unbelievable." And Arthur Caplan contended, "It would be devastating to grow up knowing you were the product of a situation in which your mother was aborted. There are many difficult things a child may have to deal with in life, but I just think we don't have any scale yet for someone to find out that they exist but their mother did not come into personhood."

Caplan and others also worried that it would be impossible to obtain a properly informed consent from the "egg donor"—the fetus—for the use of her eggs. "It seems to me that no one should be able to create a child from your eggs or your sperm without your consent," according to Caplan. But if the fetus is not a person—as Caplan admitted in the same inter-view—how can he expect *it* to have any say over the disposition of its component parts? If the fetus is not a person, than the fetal ovaries are nothing more, or less, than tissues within a woman's body. And when a woman asks a physician to remove those tissues, along with others, from her body (during an abortion), she has the right to determine whether they can can be given to, and used by, other women.

One of the few bioethicists who was not opposed to the use of fetal ovaries for the creation of life was Joseph Fletcher from the University of Virginia, who suggested that a gift from women who had undergone abortions to those who were infertile would be preferable to the current dependency on adult egg donors, who must undergo an invasive procedure. "Over all, you avoid more pain and suffering," he said. "I think that's worth thinking about, that's worth factoring into the equation."

PIG FATHERS

Sperm are not like eggs. They do not appear within the body of a male until many years after he is born. It is the production of sperm, in fact, that distinguishes men from boys at the cusp of puberty. So, you might think that a man is absolutely required for the creation of new human life.

For the time being, this is still true. But probably not for long.

Here it is a good idea to digress for a moment and review the biology of sperm cells, starting from the very beginning. Within a few weeks after conception takes place, as the cells in the embryo begin to differentiate, a special group called *primordial germ cells* is set aside for a very specific purpose. Primordial germ cells have two possible fates: they can end up as eggs or they can end up as sperm. Which way they go is determined by the signals they receive as the fetus first begins to undergo sexual differentiation.

In a female fetus, primordial germ cells are instructed to differentiate into primary oocytes, which remain quietly ensconced within the newly developed fetal ovary until, starting at puberty, they are called upon one-by-one to mature into eggs ready for fertilization.

In a male fetus, primordial germ cells are instructed to differentiate into spermatogonia, which are also sequestered, this time in the fetal testes. Starting at puberty, these cells also receive signals to restart the differentiative process.

An adult male continues to produce millions of sperm every day of his life. He is able to because his spermatogonial cells act to renew themselves during each cycle of differentiation. Each time a spermatogonial cell divides, one of its "daughter cells" retains the properties of the parental spermatogonium, while the other "daughter" is sent down the pathway of sperm differentiation. Along this pathway, which takes many weeks to complete, dramatic changes in shape and function occur, ending with the production of mature sperm cells, or spermatozoa, that are released from the testes and sent to wait in a storage area called the vas deferens until they are needed during ejaculation.

Big, round, sedentary spermatogonia cannot be coaxed into becoming sleek swimming spermatozoa in the laboratory. The process can only occur within the testes where contact between differentiating sperm cells and other types of testicular cells is undisturbed, and where still unknown signals race back and forth between each cell type. This might have appeared to be an enormous obstacle to any scientist who was interested in manipulating the process of sperm differentiation. But Ralph Brinster, a professor at the University of Pennsylvania, didn't see it that way.

Brinster was already well-known in the community of experimental embryologists for his seminal work on chimera formation (which we'll discuss in a few moments), and genetic engineering of embryonic cells (which is the focus of chapter 18). He always seemed to find a way to leap over experimental roadblocks and accomplish what had been deemed impossible.

Thus, it was not surprising that Brinster would figure out a simple way around the problem of getting sperm differentiation to work in the laboratory. What he actually did was to eliminate the laboratory, and replace it with a living testes—inside another animal—instead. In May 1996, he described the results of experiments in which he had transplanted rat spermatogonia into mouse testes. Even though the rat and mouse have been evolving apart from each other for fifteen million years, and even though the differentiation of rat sperm takes seventeen days longer than mouse sperm, and even though rat sperm look entirely different from mouse sperm, they are still able to take advantage of the mouse testes environment to undergo proper differentiation.

As Brinster told his interviewer on the BBC program "Tomorrow's World" on June 1, 1996, "The rat-to-mouse (result) suggests that you can go across species barriers; which species, it's difficult to say. In terms of going from human to mouse, it may be much more difficult than going from human to pig, and only doing the experiment will tell you."

It seems likely that a species will be found that allows the proper differentiation of human sperm from transplanted spermatogonial cells. And since spermatogonial cells are self-renewing, once an animal receives a human transplant, it will be able to produce that person's sperm for the rest of his life. Furthermore, since spermatogonial cells are easily frozen, and easily transferred from one animal to another, it would be possible for a series of animals to keep producing new sperm from the same original man for thousands of years after he has died.

Once again, the bioethicists were called upon by the media to respond to another new reprogenetic possibility. This is what Arthur Caplan had to say when Ralph Brinster's result was first announced: "Part of the way we think of who we are and how we value ourselves has to do with our origins and reproduction. Something is challenging the specialness of humanity if you originate human beings in some animal's reproductive tract."

He is absolutely right. The "specialness of humanity" has been challenged, and it's been found wanting. What this new technique, and so many others like it, tell us is that there is *nothing* special about human reproduction, nor any other aspect of human biology, save one. The specialness of humanity is found only between our ears; if you go looking for it anywhere else, you'll be disappointed.

PUTTING IT TOGETHER

What was left unspoken in the original reaction to Brinster's results was a logical extension to his technology: if you can recover spermatogonial cells from an adult, and bring them to maturation, then why not from an aborted male fetus? There is no obvious reason why conditions couldn't be found to make this work as well.

Now imagine what could happen if you combined the technology of fetal sperm maturation with the technology of fetal oocyte maturation. You could start with aborted male and female fetuses, and recover spermatogonia or primary oocytes, respectively, from each. The spermatogonia would be matured into sperm within the testes of an appropriate animal species, and the primary oocytes would be matured into fertilizable eggs within a laboratory incubator. These cells could be brought together through IVF to produce a new embryo that is introduced into the uterus of one of the women who originally experienced the miscarriage. If this pregnancy was successful, the woman could give birth to her genetic grandchild without ever having been a genetic mother. And the child itself would have two *virtual* genetic parents only, who had never come into existence on their own.

In genetic terms, fetal mating can be viewed as the opposite of cloning since it leads to the skipping of a generation between mother and child. In contrast, as noted earlier, cloning can add an extra generation (in the cloned person herself) between a progenitor parent and a genetic child. Thus, the use of either of these reprogenetic technologies will serve to confuse traditional notions of heritage and family relationships.

Fetal mating is likely to repulse most members of society, no matter what their individual views on abortion might be. And for the most part, it is hard to imagine many circumstances under which a scenario of this sort would actually come to pass. However, like every other new reproductive possibility that becomes feasible, fetal mating is sure to find a use as a solution to someone's special reproductive problem. One such use will be described at the end of the next chapter.

Shared Genetic Motherhood

All educated people know that fertilization occurs with the fusion of a single sperm and a single egg. This means that, except in the case of an identical twin parent, each child must have but one genetic mother and one genetic father. And even in the exceptional case of a twin mother or father, a child still only receives a single maternal and a single paternal contribution to his or her genetic makeup.

Many happily bonded couples view the birth of a child who brings together their genetic material as the ultimate consummation of their love for each other. And as we have seen, when barriers lie in the way of achieving this goal, many couples will do anything within their power to overcome them. A certain type of happily bonded couple, however, has never even considered the possibility of joining their genes together in a child. I am speaking, of course, of same-sex couples.

Most people think it is biologically impossible for two unrelated women (or men) to pass on their genetic material together to a single child. But by now, you know that the future possibilities of reprogenetics are almost unlimited. By now, you can probably guess that there must be some way for reprogeneticists to work their wonders and overcome the biological law that decrees only a single maternal and paternal contribution to each embryo and child. Indeed, there is a way for two women to pass on their genes to a single child, but not as you may think.

LESSONS FROM NUCLEAR TRANSPLANTATION IN THE MOUSE

Before discussing the actual procedure that can be used, I'll describe what might appear to be a more direct approach that unfortunately fails to produce the desired result. This approach can be illustrated with a thought-experiment in which you, the reader, play the role of the reprogeneticist. Let's say that two women want to be the genetic parents of a child. You wish to help them and you have, at your disposal, all the tools of a modern reprogenetics laboratory and an IVF clinic. First you review the relevant facts of biology. You know that the egg contributes half the genetic material to a developing embryo and that a fertilizing sperm cell is required to contribute the other half. You know as well that for the first twenty-four hours after fertilization, the genetic material contributed by the egg and sperm remain apart in separate spherical structures called pronuclei.

You say to yourself, "If I could only build an embryo with two pronuclei both obtained from eggs, but produced by different women, I could get a double-mothered girl child." And with your finely honed reprogenetics skills, you devise a scheme to accomplish your goal. First you recover mature eggs from each of the two women and fertilize these eggs in a petri dish with sperm provided by a donor. After waiting a few hours, you insert a tiny glass needle into a fertilized egg from the first woman, pull out the sperm-contributed pronucleus (which is readily distinguishable from the egg pronucleus), and eject it into a tiny waste bin. You next insert the needle into a fertilized egg from the second woman, and this time, you pull out the egg-contributed pronucleus instead. But rather than discarding this pronucleus, you inject it into the first egg. You would now have created a new one-cell embryo with half of its genetic material derived from one woman, the other half derived from a second woman, and no genetic father. You would introduce this double-mothered egg back into the uterus of either woman to allow it to develop to term.

This exact experiment was actually performed during the early 1980s by Davor Solter and Jim McGrath, the same pair of scientists who contributed to cloning technology. The only difference between the Solter-McGrath experiments and the hypothetical one described above is that the eggs and sperm were obtained from mice rather than humans. Still, their experimental results sent shock waves through the worldwide community of geneticists—not because they got double-mothered mice, but because they couldn't get them.

Solter and McGrath showed clearly that embryos subjected to nuclear transplantation could develop into healthy adult mice. But success was only achieved when an egg cell ended up with one sperm-derived pronucleus and one egg-derived pronucleus. Whenever any egg ended up with two pronuclei that *both* came from female parents originally, or *both* came from male parents originally, it couldn't develop properly into a live-born animal.

This result shocked geneticists because of its implication: genetic material passed into an egg from a mother must be different in some intrinsic way from genetic material passed into an egg from a sperm cell, and both versions are required for development to proceed normally.

We now know that this conclusion holds true for all mammals, including humans, and we now understand how mothers and fathers chemically modify tiny portions of egg and sperm DNA in different ways before fertilization. In order to carry out its developmental program, the human embryo requires the sperm version of one small set of genes as well as the egg version of an entirely different small set of genes because their separate modifications act to complement each other. The parental-specific modifications present in the DNA are copied from each cell to its daughter cells, and continue to persist even in the adult body. But they are all erased during the process of gamete production. Thus, irrespective of the way these special genes entered the body, they all come out looking male-like in sperm pronuclei, and femalelike in egg pronuclei. So, at least for the time being, the most straightforward strategy for producing double-mothered or double-fathered children cannot work.

MOUSE CHIMERAS

Twenty years before Solter and McGrath performed their nuclear transplantation experiments, a Polish embryologist named Kristof Tarkowski was working in Warsaw on a different approach to producing mice with two genetic mothers and fathers. Tarkowski's approach was much simpler than the one later used by Solter, and it worked from the start.

Tarkowski reasoned that if very young embryos could be separated into individual cells that could then go on to develop independently as identical twins, triplets, or quadruplets, it should be possible to reverse the process and combine multiple embryonic cells together to form a single animal. Tarkowski reasoned further that if cells originating from the same embryo could be brought together, it should also be possible to bring together cells from different embryos produced even by different mouse parents.

Tarkowski's simple method worked like a charm, and since his original publication in 1961, it has been repeated in hundreds of laboratories. When embryos produced by pairs of mice from two strains with different fur colors are merged together, the success of the protocol can be seen by simply looking at the offspring that are born. If an albino strain embryo is mixed with a dark colored one, the resulting offspring exhibit a patchwork coat with areas of dark fur alternating with areas of white.

It is important to understand what is and is not happening inside a merged embryo from two sets of parents. At the cellular level, nothing happens. Each individual cell retains its identity; no fusion between cells takes place. But, as the embryo develops, the cells derived from different parents mix together and communicate with each other as if they are all members of the same team. And when the animal is born, every tissue within it—including the brain and gonads—is a mixture of cells from the original two embryos.

Embryos and animals formed by combining cells from different fertilized eggs are called chimeras by the scientific community, in honor of the creature with this name from Greek mythology. The mythological Chimera was a fire-breathing monster resembling a lion in its head and shoulders, a goat in its middle, and a dragon behind. Although the Greek Chimera was clearly a monster, the modern-day chimeric mouse is no such thing. A lab-devised chimeric embryo and fetus proceed through development in a perfectly normal manner, and the chimeric animal that emerges looks and acts just like a normal mouse with only two possible exceptions.

The first exception occurs only when the two embryos used to produce the chimera carry genes that specify different colors of fur. As I mentioned, the animal that results will have a patchwork of the two fur colors and, on this basis alone, will be recognized immediately as a chimera. The second exception can occur when a female embryo is joined with a male embryo with potential consequences that will be discussed in the next section.

The ability to produce chimeras provides another example of a thought-experiment that eliminates any possibility of attributing individuality to the early embryo. Here's how it goes. Start with ten eight-cell embryos produced by ten different couples, and separate each into its component cells. One is left with 80 individual embryonic cells in a pool. Next, randomly select cells one by one from this large pool to obtain ten brand-new eight-cell embryos that are introduced into ten surrogate mothers for development to term. You started with ten eight-cell embryos and you ended up with ten living people, but there is no correspondence between the former and the latter.

HUMAN CHIMERAS

Now creating chimeric mice is all well and good, but how do we know that we could actually accomplish the same thing with human embryos? I could remind you that mouse, human, and all other mammalian embryos at this early stage are virtually indistinguishable from one another and will almost certainly respond to manipulations in the same way. This is the logic that Steptoe and Edwards followed originally in their decade-long quest to perfect conditions for in vitro fertilization in humans.

But I don't need to rely on this logic at all, because mother nature has already done the experiment for us. Since the 1950s, more than one hundred natural-born chimeric human beings have been identified by medical geneticists. Each of these people emerged from the fusion of two embryos that resulted from the fertilization of two eggs that had been ovulated simultaneously by their mother. We should not be surprised by this rare, but natural, process because we already know that embryos can spontaneously fall apart to form identical twins. If scientists can get two mouse embryos to stick together on contact in the lab, then the same thing should happen spontaneously on occasion in a woman's reproductive tract.

In almost all respects, a chimeric person—like a chimeric mouse—is indistinguishable from other members of its species. But, just like mice, there are two ways in which some chimeric humans can be recognized. If the two embryos that merged had genetic makeups programmed toward very different skin or hair colorations, then the chimeric person could have a patchy complexion or hair color. Among naturally born chimeric humans, this type of abnormality is rarely observed, presumably because all have only a single genetic mother and single genetic father.

The second distinction can only occur when an embryo with an XX genetic constitution merges with an embryo having an XY genetic constitution. During fetal development, the tissues that differentiate into the sex organs will be bombarded by conflicting signals. More often than not, signals from the Y chromosome predominate, and the individual develops normal, or nearly normal, male genitalia. But the gonads themselves will often develop as mixtures of ovarian and testicular tissues. In some cases, the combination of male and female signals can cause the external genitalia to develop into an intermediate configuration with an enlarged clitoris (or reduced penis) and other tissue intermediate between a scrotum and a vulva, with perhaps a shallow vagina or none at all. In fact, intersex chimeras can have genitalia across the entire range between normal female and normal male. And perhaps surprisingly, intersex chimeras can be fertile and have children, sometimes as a father, sometimes as a mother.

It is only when their genitalia are what physicians call "ambiguous" that chimeric human beings are usually detected. However, for every chimeric person identified through ambiguous genitalia, there are likely to be four or more other chimeric individuals who have gone through life unnoticed. These include essentially all of the chimeric people formed by the merger of two same-sex embryos and many intersex chimeras who have developed as normal men or women.

Would there be anything wrong with creating chimeric human beings on purpose? The obvious objection is that 50 percent would be intersex individuals with a developmental propensity toward abnormal sex organs and gender function. It would be immoral to have a child with such a high risk of developing such a serious abnormality. But with new technologies for the genetic diagnosis of embryos (which we'll discuss in chapter 17), this serious problem can be eliminated.

There are other reasons why people might be against the purposeful production of chimeric children, but I'll postpone their consideration until after a retelling of the story presented in the Prologue of Cheryl and Madelaine, the same-sex couple that wanted to combine their genes together within a child whom they could both call their own.

CHERYL AND MADELAINE'S BABY

The date is Tuesday, September 15, 2009. The city is Cambridge, Massachusetts. We are in one of the many private IVF clinics that exist in and around a metropolitan area with a well-educated and affluent population. Cheryl and Madelaine have arrived early for their appointment at the clinic, and they're both bubbling over with excitement, as well as hormones.

Cheryl is a thirty-eight-year-old theoretical physicist. Earlier in the year, Harvard University had granted her tenure, which provides job security for the rest of her life. She had been working single-mindedly toward this goal for as long as she can remember; certainly longer than her relationship with Madelaine, whom she has lived with for eight years. The quest for tenure had dominated her life, with nearly every waking hour given over to research, teaching, and other university activities. Now, with tenure in hand, Cheryl was suddenly freed to think about things—other than science—that she wanted to accomplish in her life. And the one thing that loomed larger than all the others was the desire to have and raise a child.

Madelaine is a thirty-four-year-old elementary school music teacher and a singer in a local rock band. She comes from a family of five children, and three of her brothers and sisters have already had children of their own. Aunt Madelaine adores all of her nieces and nephews, but she had resigned herself to never having children of her own. You see, Madelaine shares everything in her life with Cheryl. And although she would have loved to have had a child, she couldn't imagine raising a child unless she and Cheryl could both call it their own. But that had seemed impossible.

Cheryl was the first to raise the topic of having children in April of 2009, and over the next two months, she and Madelaine discussed their options. They considered adoption, but realized that their chances of getting an agency to consider them for a healthy child were essentially nil.

Not only were they gay, but Cheryl, in particular, was already categorized as being beyond the preferred age. They considered artificial insemination, but neither Madelaine nor Cheryl liked the idea that only one of them would be the biological mother, while the other would have no biological connection to the child at all.

Then Cheryl had lunch with a professor, and good friend—Mally Meselbert—from Harvard's biochemistry department. It was a beautiful day in early June. Throughout the main course, Mally nodded sympathetically as he listened to Cheryl's lament about her childlessness. Almost at the beginning of her monologue, Mally had thought up a technical solution that would satisfy both his colleague and her partner. But he was troubled by the potential consequences of its application to humans, and he kept his thoughts to himself as he slowly finished his meal. By the time coffee arrived, he had changed his mind. He decided that Cheryl and Madelaine had the right to make their own judgments about consequences. So he proceeded to explain how scientists working with mice, sheep, goats, and cows had perfected the technology of embryo fusion, and how occasionally, two human embryos could fuse naturally inside a woman's body, with the resulting birth of a fully viable and healthy child.

Cheryl listened in amazement until Mally had finished his long biology lesson. There was no need for Mally to be explicit. The implications were clear enough, and Cheryl asked just a single question: "Do you think we could find a fertility clinic that would be willing to work with us on this?" Mally thought for a moment. His experimental work on animals did not bring him into contact with medical practitioners in a professional way, but his wife was friendly with a very talented fertility specialist named Dr. Ricky Shapiro who operated her own reprogenetics clinic just outside the Harvard campus.

Cheryl had heard what she wanted to hear. She dropped a twenty dollar bill on the table and quickly moved toward the door. Over her shoulder, she shouted at Mally to hold the change for the next time they had lunch together. She had already decided to take a detour on the way back to her office and nearly flew into the Harvard Square T station and onto the Boston-bound train to catch Madelaine at her school between classes. Breathlessly, she raced into the school and found Madelaine in the faculty lounge. Upon catching her breath, she recounted Mally's entire biology lesson as Madelaine listened in amazement.

A summer filled with discussion, choices, and preparation is now over, and Cheryl and Madelaine wait their turn at the clinic. Finally, the receptionist motions them in. Dr. Shapiro is waiting for them in the clinic's egg retrieval and transfer room. They toss a coin. It comes up tails and Cheryl is the one to go first. She changes out of her clothes and into a standard hospital gown. Dr. Shapiro helps her onto the table and prepares her for egg retrieval. The ultrasound view of her left ovary comes onto the monitor and Dr. Shapiro smiles at the sight of lots of fluid-filled sacs, sitting on the surface, each containing a single mature egg. Dr. Shapiro goes to work—first on the left ovary and then the right—and within 15 minutes, she has recovered twenty-three beautiful eggs. These are quickly escorted into the body temperature incubator in the lab next door to await their fate. Madelaine's turn is next. This time Dr. Shapiro can only recover sixteen eggs, but she is confident that they are sufficient for the task at hand.

The time has come for fertilization, and Dr. Shapiro removes the tube containing the specially prepared sperm from the liquid nitrogen storage tank and plunges it into a small metal basin holding body temperature water. Cheryl and Madelaine have been allowed to watch the entire process, and as the sperm thaw, they recall the many hours they spent poring over the on-line Cryobank sperm donor catalogue for the sample that was best suited for them.

They had paid particular attention to the skin tone picture provided with each sample. They wanted to minimize the possibility of a patchy complexion, and so while their own complexions were not that different, they found a sperm donor whose skin tone appeared to split this small difference further. They realized, of course, that this wouldn't guarantee anything, but they figured, "Why not?" In addition, the young donor was a senior majoring in physics at MIT, with a straight-A average, who as a high school student took first prize in a state-wide contest for piano playing. Again, Cheryl and Madelaine realized nothing was guaranteed, but they were intrigued by the possibility of enhancing their separate talents together in their child.

Cheryl and Madelaine had decided on a girl. As a first step toward making this goal easier to achieve, they had asked the sperm bank to provide Dr. Shapiro with a fresh semen sample from their chosen donor. The sample had arrived two weeks earlier and was immediately placed into a machine called a flow cytometer, which separates sperm cells into two groups that are 90 percent enriched for either the X or Y chromosomes. The X-enriched sperm sample was recovered and stored frozen in liquid nitrogen for two weeks. Now the thawing process is concluded and living, swimming sperm come into view in the portion of the sample that is examined under the microscope by Dr. Shapiro's technician, and then by Cheryl and Madelaine as well.

The special sperm are drawn up into a pipette and a portion is released first into the dish containing Madelaine's eggs, and then into the dish containing Cheryl's eggs. The two dishes are covered and placed back into the darkness of the incubator. With the closing of the incubator's door, the day's activities are now over. Cheryl and Madelaine return to their home to wait patiently, miles away, as their embryos proceed slowly through fertilization and early development.

Three days later, they return to the clinic. Each properly fertilized egg has now turned into an eight-cell embryo. At this point, the embryos from both dishes are examined under the microscope and each healthy-looking one is transferred to an individually numbered compartment. Now, one by one, each embryo is held steady as a cell is plucked away for genetic diagnosis. Twenty-four samples—representing fifteen surviving embryos from Cheryl and nine from Madelaine—are sent to the molecular diagnostics lab, and four hours later, the results come back. Diagnosis was possible on eleven of Cheryl's embryos, and the results are nine females and two males. Only six of Madelaine's embryos could be diagnosed, but all are female.

Without a word being uttered, Cheryl, Madelaine, and Dr. Shapiro know that the possibility exists to create six chimeric girl-girl embryos. And silently, Dr. Shapiro goes to work. She looks at the test results to note which of the compartments on Cheryl's dish contain girl embryos. She scans the dish, picks up one, and moves it to a new dish with fresh fluid. She then moves one of Madelaine's embryos to the same dish. The two embryos are now exposed to a special chemical that dissolves their zona coats, and they are finally ready for the big event. With a gentle nudge, Dr. Shapiro pushes Cheryl's embryo into Madelaine's. On making contact, the two embryos stick together instantly. What were two living things a moment ago are now just one.

The merged embryo is given a new artificial zona coat, and set aside on the dish to await the formation of its sisters. Over the next fifteen minutes, Dr. Shapiro repeats the same delicate process five more times.

When she is finished, there are six new embryos that belong equally to Cheryl and Madelaine.

A few hours of further incubation are allowed to pass to make sure each merger has occurred successfully. Then, the time has come for the final procedure. When it is completed, there will be nothing to do but wait and hope that a pregnancy has taken hold. In advance, Cheryl and Madelaine had no way of knowing how many fused embryos would be formed. They had decided together that if only two embryos were available, they would be introduced into Cheryl's uterus. But, if there were more, it would be possible for both Cheryl and Madelaine to receive embryos in the hope that at least one would take.

After going over the available statistics on pregnancy rates achieved with the use of IVF by fertile couples, Cheryl and Madelaine decide that they will each have two embryos introduced into their wombs. They realize that they could have as many as four children as a consequence, but Dr. Shapiro assures them that this is extremely unlikely and, in any case, the number could be reduced by selective abortion, if they so desired.

For a week after their return from the clinic, Cheryl and Madelaine can do nothing but wait with building anxiety. Will either become pregnant? Will their hoped-for child be born healthy and normal? Will their family be accepted by the community in which they live? And then the first signs appear. On the same morning, Cheryl and Madelaine wake up—before dawn—with a feeling of queasiness. It is the signal they've been waiting for. The pregnancy test that each woman performs on her urine merely confirms the obvious.

But their ecstasy is now held in check by new and different fears. How many embryos are growing within them? Will a miscarriage take place? With a mixture of excitement and anxiety, they live through another three weeks before ultrasound can give them the answer to their first question. Together they return to Dr. Shapiro's clinic. This time Madelaine is the first one onto the table. The scan picks up just one little sac with a tiny beating heart. Cheryl has her turn, and again, there is but a single embryo, with a tiny beating heart.

With the results visible on the ultrasound monitor, there is palpable relief across the room. Cheryl and Madelaine quickly agree that bringing nonidentical twins into the world is probably even better than a singleton, since the two sisters will have each other to grow up with.

A month later, Cheryl and Madelaine undergo a final test to obtain confirmation that the fetus in each is really all-girl. They return to the clinic for what they hope will be the last time before their girls are born. Chorionic villus sampling, or CVS, is performed on each woman to recover cells produced by each fetus. The samples are, once again, sent to the molecular diagnostics lab. And a few hours later, the results come back. Each fetus is truly a mixture of cells from both mothers, and each is all-girl.

The next seven months pass by uneventfully. Madelaine continues in her teaching and performing careers. Cheryl continues to do research and teach as well. Cheryl is the first to go into labor. On June 1, 2010, she gives birth to a baby girl, weighing 9 pounds, 2 ounces. Cheryl and Madelaine name her Eve. And even though Eve is quite special inside, you would never know it from her looks and behavior. She is just one more precious baby on the maternity ward.

Five days later, it is Madelaine's turn. Her baby is smaller, just six pounds, eleven ounces. Cheryl and Madelaine name her Rebecca. Rebecca and Eve—destined to grow up as special sisters in a new world of repro-ductive possibilities.

AN EPILOGUE TO CHERYL AND MADELAINE'S STORY

Although I have placed the births of Rebecca and Eve a decade in the future, every technical detail could be carried out today. In a similar fash-ion, it would also be possible for two men to share a child. This could be accomplished with multiple eggs—donated by a woman—that were fertilized with sperm from each of the men. The chimeras formed subse-quently would have three genetic parents—two gene-dads and a single gene-mom. A big difference for men, of course, is that they would require the services of a surrogate mother. However, this is not an insurmount-able obstacle.

Just because a technology is feasible does not mean that it should be used, and there are various reasons why people might be against the purposeful creation of chimeric children. The first thought that comes to the minds of many is that the creation of chimeric children is weird, and the children themselves are "unnatural." As we have already seen in previ-ous discussions of reprogenetic technologies, this is a false argument. Although they are extremely rare (fewer than a thousand have been identified throughout the world), human chimeras are born naturally, and most lead normal lives, unaware of their special status.

Some people may think it is wrong for a child to be raised by gay parents. A direct response to this concern is beyond the scope of this book. Suffice it to say that hundreds of thousands of children are being raised by gay parents right now in America, and high courts in several states have acted to legitimize same-sex parenting arrangements. One cannot reject chimera formation solely because of the possibility that it would be used primarily by gay parents.

As an aside, it is interesting to note that a major argument used by the Religious Right in its opposition to same-sex unions is based on the notion that marriage is supposed to serve the purpose of procreation. According to this line of reasoning, gay unions should not be sanctioned because they are biologically barren. If we take the Religious Right at its word, the ability of gay women, or gay men, to co-procreate should validate their right to become married.

Some people contend that medical treatments should only be used for curing diseases, and not for frivolous activities such as chimera formation. In American society, at least, this argument fails because we generally accept the right of people to use medical services for other frivolous purposes, including facelifts and tummy tucks. As a society, we accept the right of people to use medical services for whatever purpose they wish, so long as they pay for it themselves; that's the American way.

The most serious argument is that chimera formation might be harmful, in some direct way, to the child that emerges. People of all political and religious persuasions would almost certainly consider it immoral to perform a reprogenetic protocol that would cause a child to be brought into the world with a defect that could impinge significantly on that child's health or happiness.

Yet in terms of general health characteristics, vitality, and life span, chimeric children will appear no different from any other children. We know this is true from evidence accumulated in studies of thousands of chimeric mice as well as the handful of chimeric people who have been identified and examined. In terms of brain function as well, human chimeras appear to be perfectly normal. Of course, subtle problems might be hidden in the small number of humans examined, but based on our understanding of development, there is no reason to expect any.

Harm can be incurred psychologically as well as physically. There are sure to be some who will argue that even if they look and think normally, chimeric children will "feel bad" because they know they're chimeras. It seems hypocritical for anyone willing to deny government-sponsored food and shelter to children born into poverty to make this argument. Indeed, the same false argument has been used against IVF, artificial insemination, egg donation, and surrogacy. Millions of prospective parents have not found it persuasive.

The physical attributes that might distinguish chimeric children from other children are skin complexion and hair color. With current technology, it is not possible to assess fully the risk, or prevent the occurrence, of a patchy complexion or head of hair. Although not a medical problem *per se*, this difference could impinge upon the happiness of a chimeric child. Whether the severity of this problem is serious enough, and the risk of its occurrence large enough, to argue against this use of this technology is not clear. Hair color could be homogenized with dyes. But, skin color patchiness might be more difficult to hide and could be viewed as unattractive by others.

The possibility that a chimeric child could be born with a permanent patchy complexion and hair color might be enough to dissuade some same-sex couples from using this approach to co-procreation. If so, there is an alternative approach that does not depend on chimerism, although it could raise problems of a different sort. It is based on the idea of passing one woman's genetic material through a "testicular filter" to enable it to come together in the same embryonic cells with the genetic material from a second woman. Here's how it would work for Cheryl and Madelaine:

Cheryl would use IVF with donor sperm enriched for the Y chromosome to become pregnant with a male fetus. Similarly, Madelaine would use X-enriched sperm to become pregnant with a female fetus. Both women would undergo induced abortions near the end of their first trimester, and spermatogonia and oocytes would be recovered from the two fetuses and matured into sperm and fertilizable eggs, respectively. These would be combined through IVF to produce embryos that were frozen in liquid nitrogen until Cheryl and Madelaine were ready to begin new pregnancies. Then the embryos would be thawed and introduced into both, to enable the birth of shared children nine months later.

The children that emerged from this scenario would be related equally to Cheryl and Madelaine. Furthermore, since they would each be derived from a typical union between a single sperm and a single egg, there would be no risk of skin color patchiness of the type that might occur in chimeras. Strictly speaking, of course, the children would be the genetic *grandchildren* of Cheryl and Madelaine. But this technical detail should not make any difference. With both approaches to co-procreation—chimerism or fetal mating—a child receives 25 percent of its genetic material from each woman, and the other 50 percent from male donors.

While the fetal mating approach eliminates the potential problems associated with chimerism, it raises the specter of purposely killing fetuses as a means toward the creation of human life. Although late first trimester abortions have been sanctioned by the Supreme Court, and the fetus at this stage does not have sentience, it does look like a little human being. Even those who are adamantly pro-choice may feel some revulsion against the notion of fetal mating. But, there is still some time, perhaps, to think about this all, because in 1997, the maturation of fetal spermatogonia in animal testes has yet to be accomplished. On the other hand, the original chimera scenario described for Cheryl and Madelaine could be carried out today.

16

Could a Father Be a Mother?

The inch-and-a-half-high headline at the top of page one read "MAN PREGNANT." Just beneath this was the three-quarter-inch subheadline: "HE BECOMES SURROGATE MOM FOR STERILE WIFE." The date was September 29, 1987, and the "news" medium was the *Sun*, an American supermarket tabloid.

It seems that a sterile Finnish woman named Mauna Koinevo wanted to have children, but couldn't accept the idea of using an unrelated surrogate mother. In Mauna's words, "I knew I couldn't bear children, but then I heard about work being done to impregnate men." She convinced her husband Richard to inquire about this work at a nearby "research clinic" in Vaasa, Finland. There he found Dr. Kolavo Sarvast. The good Dr. Sarvast agreed to help the Koinevo family achieve their goal. "We had been working on a method to allow men to become pregnant and were looking for volunteers," he said. "Richard seemed an ideal candidate."

Four years later, on April 30, 1991, the very same newspaper ran what they called the "Story of the Century." This time the page one headline read "PREGNANT MAN GIVES BIRTH—BABY LIVES!" The pregnant man was named Giovanni DiPenza, and the place was Palermo, Sicily. The story quoted Giovanni's doctors as saying, "The precedent-shattering operation marked the first time that a man ever gave birth to a live baby." One wonders how the editors at the *Sun* conveniently forgot about Richard Koinevo.

The fantasy of a man becoming pregnant is probably as old as storytelling itself. But fantasy it clearly was until the perfection of IVF in 1978. IVF, of course, would be an essential ingredient in any attempt at male pregnancy. And with this obstacle overcome, reproductive biologists began to ponder whether the unimaginable just might be . . . imaginable.

From the early 1980s onward, there have been sporadic reports of scientists musing aloud about the possibility of maintaining a pregnancy within the abdomen of a man. And in 1985, *Omni* magazine published the first full-length popular article on the topic—complete with technical details on how to make it work—by Dick Teresi.

By the beginning of 1995, there may still have been a few Americans who had not read or heard talk about male pregnancy, but that was soon to change with the release of the motion picture *Junior,* starring Arnold Schwarzenegger and Danny DeVito, playing modern-day equivalents of the pioneering IVF scientist-physician team of Edwards and Steptoe. For reasons that are highly contrived, the DeVito and Schwarzenegger characters decide they must perform an experiment in reproductive biology on themselves. This leads the DeVito character to carry out IVF with a donor egg and the Schwarzenegger character's sperm. He then uses ultrasound to find a good location in Schwarzenegger's peritoneal cavity to place the fertilized egg for implantation. A week later, he reads the positive pregnancy test and wryly announces, "You may be crazy, but you're also pregnant."

With the daily ingestion of hormones, the Schwarzenegger character is able to maintain his pregnancy. And in one scene with not-too-subtle political undertones, he flees the clutches of a villainous administrator, while exclaiming, "It's my body, my choice." The movie heads toward its climax with the delivery, by modified Cesarian section, of a healthy baby girl. In the final scene, a hymn to family values, the baby, the gene-dad/birth-mom, and the egg-donor/gene-mom (who falls in love with the Schwarzenegger character after he becomes pregnant) play happily together on a California beach.

Movies and novels that mix real science with science fiction often lead to confusion in the public mind about what is really possible and what is not. Usually, scientists and physicians can be counted on to sort it all out. With male pregnancy, though, something funny happens. Some say it is possible while others say it isn't. To understand how different professionals can reach such opposite conclusions, we must delve into the thought processes of the scientist and the clinician, respectively.

The scientist lists the ingredients that are essential for starting and maintaining a human pregnancy. The first ingredient is a fertilized egg. The second is an appropriate hormonal environment to allow implantation and pregnancy to proceed. The third is a living womb within which the embryo can implant and form a placenta. All three ingredients occur naturally in a fertile woman. Could they be duplicated, as well, within a man?

First, let's consider the fertilized egg. The birth of Louise Brown in 1978 demonstrated that these microscopic entities were no longer the exclusive province of fertile women. Eggs fertilized in vitro could be picked up through a tiny glass needle and placed anywhere, including the space inside a man's body.

Without the proper hormonal environment, however, implantation and pregnancy cannot take hold. So it is critical to know whether a pregnancy hormonal environment could be replicated in a man's body over the entire nine months of gestation. You might be surprised to learn that we already know the answer to this question, and it's yes.

After a woman passes through menopause, she is no longer able to produce the hormonal environment associated with pregnancy. And yet, with the use of donated eggs and IVF, postmenopausal women in their fifties and sixties have gotten pregnant and given birth. The pregnancies were achieved and maintained through hormonal injections that simulated the pregnancy environment of a naturally fertile woman. Based on our current understanding of endocrinology, there is no reason why the same type of pregnancy environment could not be simulated in a man as well.

"Eggs and hormones are all well and good," you might say. "But, it's the third ingredient that can't be duplicated. Women have wombs and men do not. And that's always the way it's going to be."

Once again, mother nature tells us that our intuition is faulty. Every once in a while—in one pregnancy out of ten thousand—the fertilized egg doesn't make it to the uterus, and ends up instead in the wide open space of the abdomen, also known as the peritoneal cavity. This can happen because the ovary is not actually attached to the fallopian tube (or oviduct), as is commonly thought. Instead, after ovulation, the egg must make its way into the nearby opening at the end of the tube in order to begin its journey toward the uterus. Occasionally, when conception occurs very close to the opening in certain women, the newly fertilized egg may actually fall back out of the tube and into the abdomen.

Now you might think that once an egg has fallen into the abdomen, its chances of survival are nil. Surprisingly, at the appropriate time of development, an embryo can implant itself into almost any living tissue that it happens to alight upon. And the abdomen is filled with all sorts of tissues—from the intestines to the kidneys, to the liver and the spleen. With successful implantation and sufficient placental formation, the embryo can develop normally into a fetus that can be carried through a full nine months of pregnancy. At the end, of course, it has nowhere to go unless it's delivered by a modified Cesarian section. The medical literature is filled with sporadic reports of healthy live-born babies that were carried by mothers pregnant in this unusual way.

So let's come back to the third ingredient required for pregnancy: a living womb within which the embryo can implant and attach a placenta. If a woman's abdomen can act as womb, a man's abdomen could do just as well. "Clearly," the scientist would conclude, "I've now proven that human male pregnancy is possible, and it's possible today!"

"Wait just a minute," the clinician would implore, "let's look at all of the reported cases of abdominal pregnancy again, this time with a greater eye to the clinical details. And let's start out with some of the general statements made by the reporting physicians":

"Abdominal pregnancy is a rare but life-threatening condition."

"Morbidity and mortality for both the fetus and the mother are considerable . . . Once the diagnosis is established, immediate surgical intervention is usually advisable."

"Care of the patient afflicted with it may present formidable challenges."

Abdominal pregnancy is considered "life-threatening" because of the placental connection the embryo must set up between itself and the body within which it lies. In a normal pregnancy, the embryo attaches itself to the specialized internal lining of the uterus known as the endometrium. Endometrial cells are recruited along with embryonic cells to form the placenta, but at the time of birth, the entire placenta detaches itself easily from the intact uterine wall to follow the baby through the birth canal. The ability to create a detachable endometrial lining that can be incorporated into the growing placenta is a unique property of the uterus.

Unfortunately, when an embryo implants into an abdominal tissue, detachment is not so simple. The problem is that the development of the placenta can cause complete intermixing between embryonic and host tissues so that there is no clean boundary between the two. The more extensive the intermixing is, the more problematic it becomes to remove placental tissue. The physician has to cut between the wholly placental tissue, and the intermingled placental-maternal tissue. Large blood vessels must be severed, and as a consequence, difficult-to-control internal bleeding can take place.

Problems are not just confined to the stage at which a pregnancy is terminated. Long before the final event, a placenta can cause severe damage to an organ that it's invaded, with the possibility of spontaneous hemorrhaging that can quickly result in death.

So is male pregnancy possible? Probably yes.

Is male pregnancy feasible? No, not at the present time. It's not just a question of whether the "Baby Lives," as the tabloid writer for the *Sun* thought, but whether the pregnant man himself survives the gestation and birth.

Today, at least, the attainment of pregnancy is not something that any sane man would attempt, or that any ethical physician would suggest. Once again, though, we should never say never. At some point in the future, it's likely that reproductive biologists will figure out how to direct the growth of the placenta away from vulnerable abdominal organs and onto an easily detachable but blood-rich surface for growth.

When male pregnancy becomes available, we might want to ask whether any man would really want to put himself through it. As the DeVito character in the movie *Junior* says, "Part of the beauty of being a guy is not having to get pregnant."

While most guys may feel this way, there are certainly a significant number who do not. These will include some transsexuals—people born physically as men who undergo sex change operations to become women later in life—as well as married men who want to be surrogates for their infertile wives, just like the *Sun*'s Richard Koinevo, the birth-mother-gene-father from Vassa, Finland.

As we think about all the variations on motherhood and fatherhood described in this part of the book, it is clear that these concepts are not as easily defined as they once were when human reproduction was a mysterious process hidden from the view of all, within the womb of a woman. In those bygone days, a child had but one mother and one father, and that was that. But with the unveiling of human conception—literally in broad daylight—reprogeneticists gained the ability to finesse the single passageway that had always connected parents to their children. Today, there is not one but many paths that can be followed to reach the goal of having a child. The validity of any of these paths should be judged not by their intrinsic nature, but by the love that a parent gives to the child after she or he is born.

PART FIVE

TOMORROW'S CHILDREN

For God knows that in the day you eat
of it [the fruit of the tree of knowledge]
your eyes will be opened, and *you will be
like God*, knowing good and evil.

—GENESIS 3:5

17

The Virtual Child

ALICE

At the top of the window on the computer screen is the smiling image of a teenage girl, about sixteen years old. Alice has a round face, small lips, a slightly broad nose, small ears, and hazel eyes just like those of her maternal grandmother and an uncle on her father's side of the family. Her thick, dark brown, wavy hair is on the oily side but otherwise nondescript. Traditional vital statistics are displayed to the right of her image: sex— female; height—between 5 feet 3 inches and 5 feet 5 inches; weight— between 122 and 129 pounds. And at the bottom of the window is a list of general categories that contain more specific information on who this Alice is, and what she can expect from life. The categories are Severe Single Gene Disorders; Predispositions to Complex and Infectious Diseases; Physiological and Physical Characteristics; and finally, Innate Personality and Cognitive Characteristics.

Melissa slowly takes hold of the mouse and tenatively moves it until the arrow on the screen points to category number one—"Severe Single Gene Disorders" or SSGD. A single click opens up a new window with a long list of several thousand diseases, any one of which would have devastating effects. Included are sickle cell anemia, cystic fibrosis, Tay-Sachs, and PKU, which each afflict one in every thousand or so children. But the vast majority of diseases on the list are exceedingly rare. In the column next to each disease name is one of three symbols—a green + to indicate the presence of two good copies of the gene; a yellow *C* to indicate carrier (but not disease) status, with one good and one mutant copy of the gene; or a bright red *D* to indicate the presence of two mutant copies of the gene and disease inevitability.

Summary information can be viewed at the top of the list. This Alice has tested green on 4,234 diseases; these are the ones for which she is neither at risk nor a carrier. She tests yellow—carrier status—on six very rare diseases (the average person tests yellow for eight), and there are no diseases for which she tests red.

The SSGD window is closed, and a click on "Predispositions to Complex and Infectious Diseases" or PCID brings up another window with another list containing several thousand entries. Each disease listed in this category results from the interaction of genetic and environmental factors. In almost all cases, the genetic contribution is spread out over multiple genes. In the column next to each disease name is a number between 0 and 100 that describes the inborn risk factor in comparison to the general population. Numbers below 50 represent decreasing risk relative to the average, numbers above 50 increasing risk.

Risk of childhood leukemia is 45, slightly below average. The long-term risk of breast cancer is 55, which is slightly above average; this means that by the time she reaches the age of 55, this Alice will have a 6 percent rather than a 3 percent chance of being so afflicted. Scrolling down the long list shows that most risk factors lie within the average range, while a handful are somewhat more extreme. All risk factors above 65 are highlighted in red.

One red-highlighted entry is heart disease, with a risk factor of 70. Further relevant information is obtained by a click on the disease term, which brings a smaller window into the foreground. The text in the small window points out that this Alice's specific heart disease risk factor is responsive to environmental modifications. The 70 risk factor value is based on the average American diet and the average American exercise routine. But with the changes in lifestyle that are specified, risk of heart disease can be reduced to 53—almost average. The other highlighted, high-risk entries are also amenable to lifestyle changes or are of a nature that is not deemed serious.

A few more clicks and all of the PCID windows are closed. Now Melissa moves down to the "Physiological and Physical Characteristics" category, and brings up a list of various bodily attributes and measurements that can affect general health as well as athletic talent. An overall assessment shows Alice to be generally healthy. One minor concern is that her skin is sensitive to sunlight, but with the judicious use of sun blockers, she'll be able to avoid any ill effects. As far as athletic talent goes, there's not much out of the ordinary for a girl of her size and body shape. If she had the desire, and she pushed herself quite hard, she might be able to make her high school gymnastics team.

Melissa returns to the main window with the face of the girl herself. She focuses her eyes and her mind on the image, and whispers, just under her breath, in an almost quizzical way: "Alice? . . . Alice?"

She pauses for a moment, then closes the window on this Alice, number 17, which brings her back to the master list. She scrolls down the list to number 43, clicks once, and a new window opens up to a new girl, with a facial appearance that is quite different from the first Alice. This girl has a longer face with higher cheekbones, brown eyes, and straight dirty blond hair. Melissa focuses in on this second image with the same mantra as before: "Alice? . . . Alice?"

Which one will become the real Alice? It has now come down to just two of the eighty-four available in the original list. Eighty-four genetic profiles obtained for eighty-four embryos sitting safely and quietly at −320 degrees, while Melissa and her husband Curtis make up their mind. The choice of a genetic profile for their child is the most important decision they will make in their lives and there is no reason to rush it.

The first cut had been easy. Twenty-seven of the original ninety-six profiles showed chromosomal abnormalities or other mutations that were incompatible with normal development; had any of these embryos been implanted, a miscarriage would have ensued. The remaining sixty-nine had the potential to go full term, but six of these showed evidence of severe genetic defects that were likely to cause death during childhood or young adulthood. Another twenty-two had disease predispositions, physical or cognitive starting points that were below average, or extreme aspects of one temperament or another that Melissa or Curtis deemed undesirable. Of the forty-one profiles that remained after this cut, twenty-three were female and eighteen were male. Melissa and Curtis had already decided to have a girl, and they had already decided to name her Alice. This decision meant that they had only twenty-three profiles to consider in greater detail . . . twenty-three potential Alices.

Before they started the process, Melissa and Curtis knew that the genetic constitution of their child would be limited by their own genetic profiles. They knew that they could not expect a blue-eyed baby with light blond hair, or a child with the physical attributes required to become a star tennis or basketball player. But a number of other positive characteristics were within the range of possibility. And what Melissa wanted most was a daughter with an innate score for musical talent that was even higher than her own high score. Curtis's hopes were focused elsewhere. He wanted Alice to be born with temperament and cognitive attributes that would serve her well in the business world.

Melissa and Curtis also know that there is no such thing as a perfect child. From the time that complete genetic profiles became a reality, every adult who had one done could see the combination of flaws and foibles that were uniquely their own. And when it came time to have a child, it was up to the parents to decide which should be avoided and which were tolerable.

Each one of the twenty-three potential Alices came with strong points and weak points, in the eyes of Melissa and Curtis. The one with the strongest musical talent was on the shy side and lacking in talent for abstract analysis, which did not appeal to Curtis. Likewise, the one with the strongest talent in abstract analysis would have a tendency to be twenty pounds overweight and not very musical, which Melissa was unwilling to accept. So compromise was the word of the day.

And that is how they came down to the final two. Both scored well on all important temperament scales with a tendency toward long-term happiness, emotional stability, conscientiousness, mild risk-taking behavior, and an assertive but not overaggressive nature together with an outgoing personality. In addition, both had well-above-average talents in both music and abstract analysis. Now the decision came down to small aspects of physical appearance. Number forty-three had a chin, nose, and eyes that made her resemble her father, and this appealed to Curtis. But Melissa thought that number seventeen would turn out to be more pretty, although this highly subjective aspect of physical appearance would always be difficult to gauge in a computer-generated facial image. Melissa and Curtis finally made their choice, and nine months later, they shared in the thrill of the birth of their baby girl—a real-life Alice.

WHERE THE TECHNOLOGY NOW STANDS

The concept of virtual children became reality on April 19, 1990, with a report in the journal *Nature* of two pregnancies that had been established with individual embryos chosen on the basis of their genetic profiles. The potential mothers, who had volunteered for this first clinical test of what is referred to as "embryo biopsy" or "preimplantation genetic diagnosis" (PGD), were all carriers of a serious disease mutation that could be expressed in sons but not daughters. Thus, mothers could be certain their children would be disease-free so long as they could be certain that their children would be girls.

In the first use of PGD technology, it was only necessary to determine whether a single, arbitrary piece of DNA on the Y chromosome was present in—or absent from—each embryo. Its presence demonstrated the maleness of an embryo; its absence was indicative of femaleness. Only embryos that scored as virtual girls were chosen for implantation. Nine months later, their mothers gave birth to real girl babies who would never express the family-borne disease.

Now in 1997, it is possible to screen thousands of different genes within individual embryos to determine various ways in which the virtual children associated with these embryos differ from one another. To appreciate what it means to screen a gene, we must first understand how people differ from one another genetically.

Although it is commonly said that someone "has the sickle cell gene" or a "gene for red hair," or "the breast cancer gene," this kind of talk does not present an accurate picture of what's happening at the genetic level. In fact, we all share the same set of 100,000 genes in two copies laid out along 23 pairs of chromosomes. The total of all the information in these chromosomes is referred to generically as the "human genome." No human being alive (at the time of this writing) has any extra genes. There is no sickle cell gene or red hair gene or breast cancer gene. There are only alternative forms of genes that we all share. It is different *forms* of the same genes, rather than different genes, that distinguish us from one another.

The alternative forms of a gene are called *alleles*. Most alleles of a single gene differ from one another in a small way, often by as little as a single DNA unit known as a *base*. A DNA base is analogous to a *bit*, which is the basic unit of information stored in computers. Just as a computer file is composed of a series of bits (arranged in bytes), each gene is composed of a series of hundreds or thousands of bases. But even a single base difference in a gene can have drastic consequences. One gene that we all carry is called beta-globin. A particular single base change in this gene alters the structure of the hemoglobin protein, which leads to the devastating disease of sickle cell anemia.

Mutations—changes in bases that lead to new alleles—are not always bad. We are all children of mutants. In fact, every one of our genes has been built up through an evolutionary process that involved one mutation after another. We tend not to think of ourselves in this way because the term "mutant" has such a negative connotation, but it is true nonetheless.

Once different alleles have been identified, it becomes possible to screen for their presence or absence within individual embryos. Screening has already been performed for diseases such as sickle cell anemia, cystic fibrosis, Tay-Sachs, and Huntington Disease, which are each caused by a mutant allele in a single gene. The same technology could be used just as readily to detect alleles associated with positive traits associated with better health, temperament, or talent. As long as a DNA difference has been identified between two alleles at a gene, that difference can be detected with the techniques of modern molecular biology.

But how, you may ask, is it possible to analyze individual genes in single cells plucked from an embryo? When you understand what actually has to be looked at, you'll see what the problem is. Single cells carry only a single molecule of DNA with the instructions for each gene copy, and the difference between alleles can be confined to just a dozen atoms. This means that if you want to know whether a particular allele is present in an embryo, you must have a technique that can distinguish whether a particular set of twelve invisible atoms—hidden among the trillions of atoms that make up an embryonic cell—are present in one position or another. The technique must be rapid, accurate, cheap, and easy to perform on large numbers of samples.

There is no chemical technique that can provide information on the atomic structure of a single molecule. It was for this reason that scientists assumed, before 1983, that it would always be impossible—not just unlikely, but absolutely impossible—to perform a genetic diagnosis on a preimplantation embryo. But these preconceived notions of scientific limitations were completely erased with the invention of the Polymerase Chain Reaction or PCR.

PCR was conceived of by a single eccentric scientist named Kary Mullis in a manner that has already become a legend in what is still a very young field. This is the way Jim Dwyer of the New York *Daily News* described it based on his interview with Mullis in 1993:

> Mullis was being playful on an April evening in 1983 as he drove up to his ranch. "My hands were occupied, but my mind was free," he says. He remembers the fragrance of the flowering roadside buckeye that washed in the car windows as its white stalks bobbed in the headlights. How, he was pondering, could you find a single spot on the long, fragile DNA molecule? In a series of acrobatic chemical leaps, he realized that a section of DNA containing a gene or fragment could be marked off, then forced into copying itself using replicating techniques similar to those DNA employs when a cell divides. Then he realized something so startling he had to pull the car to the side of the road. When he had been messing with computer programs, he had been impressed by the power of a reiterative computer loop, in which the same process is repeated over and over. He saw how fast numbers can climb when they increase exponentially. Replicating DNA could work about the same way: By adding the right chemicals, the little section of DNA could keep reproducing itself automatically and exponentially—so that the fragment would double, from two pieces, to four, to eight . . . and ever onward. In practical terms, he saw that, after 8 doublings, he would have 256 copies of the gene. By the twentieth cycle, he'd have 1,048,576. By the thirtieth, he'd be up to 1,073,741,824—a billion copies of a single gene in three hours.

Although the full ramifications of PCR would later be elucidated by others, Kary Mullis had invented—in his mind that night—a method for beginning the workday with *a single cell* and ending up with a test tube full of billions of copies of a single gene before lunch. Three hours to clone a gene by a machine that can be purchased for less than $5,000 (and is now present in high school science labs across the country), and starting with just a single DNA molecule. DNA alleles could be read after lunch, and within the course of a single workday, genetic profiles could be ready on hundreds of different embryo samples.

More than any other technique invented during the twentieth century, PCR has changed the course of the biological and biomedical sciences. In addition to the enormous power that it added to gene discovery and analysis—it now plays a role in nearly every experiment performed in every genetics laboratory in the world—PCR has made it possible to obtain rapid genetic profiles not only on humans but other animals and plants as well, with an enormous impact on both agriculture and environmental science. PCR has also had an enormous impact on forensics with its power to provide genetic profiles on even single hairs left behind at the scene of a crime. And PCR has provided us with the ability to look back into the past, to demonstrate that the skeletons found buried in an isolated Siberian town really did belong to the last Russian Czar and his family, and much further back to derive genetic profiles on insects and plants that have been extinct for millions of years. The real-world recovery and analysis of DNA from Jurassic-age bugs trapped in amber was the premise on which *Jurassic Park* is based. In recognition of the enormous impact that his road-trip discovery has had on many areas of human endeavor, Kary Mullis was awarded the Nobel Prize in chemistry in 1993.

We are finally ready to follow the entire process of preimplantation genetic diagnosis (PGD) from start to finish, as it is accomplished today. First, a woman is stimulated hormonally to ovulate a large number of eggs—typically a dozen or so, but occasionally up to thirty. These mature eggs are recovered from the ovaries and placed in a petri dish where they are incubated with sperm to achieve fertilization. The new embryos are allowed to develop in an incubator for two and a half days, at which point each one contains between six and ten cells. A chemical drill is used to make a small hole in the zona coat surrounding each embryo, and a microscopic needle is passed into the hole to extract one or two embryonic cells. These cells—and the DNA molecules that they contain—are dissolved in a solution, and the PCR protocol is used to amplify predetermined regions of the genome a billion or more times. Other molecular technologies are used to determine which alleles are present, and the information obtained allows the selection of particular embryos for introduction back into a woman's reproductive tract.

It is time to point out that significant limitations in the practice of PGD still exist. First, it is usually not possible to obtain more than a dozen eggs from a woman after hormonal treatment. Since not all eggs undergo fertilization and proper development in a petri dish, the number of embryos available for analysis is reduced further. And even if an embryo develops normally and is successfully biopsied, the rate of success in retrieving genetic information for even a single gene within it is only 90 percent. Based on this percentage, simple probability rules tell us that the chance of successfully determining a profile for just 10 genes in a single embryo is only 35 percent. In other words, if an average couple wants to select a child based on a genetic profile of just ten genes, it will probably have less than four embryos (35% × 12) to choose from. To make matters worse, the probability that any introduced embryo will actually implant into a woman's uterus is still less than 50 percent under the best of circumstances. This situation means that at the end of the process, there is a good chance that no child will emerge.

All these limitations constrain the utility of the current PGD technology to just those couples who are at known risk of having a child afflicted with a disease caused by the presence of a single disease allele in one or both prospective parents. In this context, PGD can be used to separate those embryos that will not be so afflicted for an attempt at pregnancy.

PGD is greatly appreciated by the small number of couples for whom it is now useful. At the moment, though, the Alice scenario that introduced this chapter remains securely in the realm of fiction. But for how long?

WHERE SCIENCE AND TECHNOLOGY CAN GO

There are five technical problems that must be solved—along with a scientific database that must be established—before the Alice scenario becomes reality. First, the efficiency with which genetic information is obtained from any individual gene—in each embryonic cell analyzed—must be increased essentially to 100 percent. Second, all 100,000 human genes must be identified along with the common alleles at each that are carried by most people. Third, a method must be devised for screening all of these genes rapidly and efficiently. The data obtained from such a screen would provide a complete genetic profile for each individual embryo. At this point, a scientific database must be available for matching each genetic profile with a description of a corresponding virtual child. Fourth, a method must be developed for increasing the number of eggs that can be obtained for IVF from any individual woman to at least one hundred. Finally, the efficiency with which any chosen embryo can be turned into a baby must be increased to 90 percent or greater.

Incredibly, solutions to all five technical problems can be imagined based on technologies available to us right now. Almost surely as well, the required scientific database will expand rapidly to completion within the next half century, if not sooner. Let's examine the potential solutions one by one.

The first problem concerns the efficiency with which genetic information can be obtained from the single copies of each DNA molecule present in the single embryonic cell removed for genetic diagnosis. Although PCR is powerful, it is unlikely that its efficiency can be increased to anywhere near the levels required for recovering information from all 100,000 genes present in a single cell. There is, however, an alternative approach that will solve the problem more readily. Instead of using chemistry to make copies of DNA molecules (the PCR way), scientists can use biology.

The biological approach is to use special nutrients and signals that will force each individual embryonic cell to grow into a larger mass of thousands of identical cells. Of course, with each cell duplication event, all the DNA also duplicates naturally. With 1,000 cells, there are 1,000 exact copies of each allele at each gene; with 100,000 cells, there are 100,000 copies. With a sufficient amount of DNA, it becomes possible to screen for the presence of alleles at any *known* gene with very high efficiency.

So how many human genes are known? In 1997, fewer than 10,000 have been fully characterized. But the situation is changing rapidly as a result of the Human Genome Project, which was initiated by the National Institutes of Health in 1991, with the goal of identifying and characterizing all 100,000 human genes. It seems likely the goal will be achieved by the year 2020. And by 2030, it is likely that all the common alleles present at each human gene in different members of the population will also be known.

For the purposes of PGD, all of this information wouldn't do us much good without a rapid and efficient way to screen all 100,00 genes in every single embryo. With the best genetic technology available in the past (meaning before 1996), it would have taken years to accomplish this task. Thus, it would not have been feasible to obtain routinely complete genetic profiles.

But the future is now. By bringing together the technology used to manufacture computer chips with chemical methods developed for the synthesis of DNA, a small biotech firm named Affymetrix has pioneered the development of DNA chips that are set to revolutionize the practice of genetics in the twenty-first century. DNA chips carry separate DNA fragments in a checkerboard pattern of microscopic blocks. Each separate block can act as a detector for the presence or absence of a particular allele. To perform an analysis, one exposes the chip to a solution containing a DNA sample, and then uses a microscope-based detection system in coordination with computer software to obtain a readout.

A 1-inch square DNA chip has already been developed with the capacity to hold 400,000 independent DNA fragments—enough to screen an average of 4 alleles at each of the 100,000 genes within the human genome. The capacity of future DNA chips is poised to climb rapidly (their inventor, Stephen Fodor, sees capacity doubling every 18 months just like computer chips), and their cost will decline rapidly as they are put into mass production. With DNA chips, complete genetic profiles could be obtained for any number of embryos within a few hours.

Even when we gain the ability to *obtain* complete genetic profiles, much of the information within them will remain meaningless. Proper interpretation of an entire profile will depend upon a deep scientific understanding of the connection between all of the many alleles that exist among human genomes and the characteristics that they control or affect. These connections are being made one by one, even as you read this book. The use of DNA chips in conjunction with information obtained from the Human Genome Project will rapidly accelerate the process so that computer generated attribute profiles—of the type described for Alice—should become feasible by the middle of the twenty-first century.

Obtaining and interpreting complete genetic profiles won't be of much use to potential parents if they can only look at a dozen or fewer embryos. This point was implicit in the Alice scenario, where a large starting pool was essential to provide a sufficient number of different genetic profiles to choose from.

The solution to this problem will probably not come from further optimization of protocols for hormonal stimulation of ovulation. Instead, a very different approach will be pursued. Girl babies are born with a million immature eggs in their ovaries. Very few of these eggs are ever ovulated during the fertile period of a woman's life. Throughout this period, egg degeneration occurs constantly, but even at a point just before menopause, there are still tens of thousands of viable eggs that remain. A very small piece of an ovary taken from a young woman will contain hundreds or thousands of eggs. Based on the technology described in chapter 14, it will soon be possible to induce most of these eggs to mature under laboratory conditions. Upon in vitro fertilization, the eggs could be turned into hundreds of newly developing embryos, each one ready to be profiled by the methods outlined above.

There is a final technical problem that must be solved before the Alice scenario comes true. If a couple has gone to the trouble of identifying an embryo to grow into their child, they won't be satisfied—at this late stage—with just a 50 percent chance of success, which is the best they could hope for today. It is possible that optimization of current protocols could increase the rate of implantation further, but probably never to the high levels required to warrant the use of the technology in the way I described for Alice. Again, a radically different path might be taken to reach the goal of an implantation rate close to 100 percent.

The trick would be to clone the chosen embryo into a mass of a thousand or more identical cells. This mass would provide an essentially unlimited number of nuclei for transfer into the cytoplasms of nuclear-free unfertilized eggs recovered from the same small piece of ovary that was previously provided by the potential mother for the production of the initial embryos used for screening. As noted previously, nuclear transfer from embryonic cells has already been successful in many species, including another primate—the Rhesus monkey. The monkey result, in particular, tells us that the same technique will almost certainly work with human embryos. With this cloning-based approach, an embryo with the same genetic profile could be introduced into a potential mother's uterus each month until a successful pregnancy is achieved.

What I have presented are potential solutions to all of the technical problems that currently limit the extent to which preimplantation genetic diagnosis can be carried out. Whether these approaches—or others based on the invention of new technologies we can't yet imagine—are actually used is not important. What matters is the almost certainty with which a world of virtual children and genetic choice will become feasible by the middle of the twenty-first century.

Again, just because a technology *can* be developed and used, does not mean that it *will* be or *should* be developed and used. And the arguments over embryo selection are likely to become louder and more fierce as the technology grows ever more powerful.

AN INTERLUDE: SOME VOICES FROM SOCIETY

These women . . . want nothing more than to be mothers; their husbands are equally fervent about being fathers. Their dreams are not of just any child. They want a child or children of their flesh, *a child with the father's chin and the mother's knack for mental arithmetic.*

—BARBARA STEWART, *NEW YORK TIMES*

Do we want to have a society where parents can flip through a DNA catalogue and design their own "boutique baby"? Will we accept that it is perfectly reasonable to discriminate against people before they are born, or prevent them from being born, because we don't like their genes.

—DEAN HAMER, GENETICIST AT THE NATIONAL CANCER INSTITUTE WHO DISCOVERED THE "GAY GENE"

Why is it OK for people to choose the best house, the best schools, the best surgeon, the best car, but not try to have the best baby possible?

—PARENT OF A CHILD BORN AS A RESULT OF ARTIFICIAL INSEMINATION WITH SPERM OBTAINED FROM A DONOR SELECTED FOR "HIGH INTELLIGENCE"

For some, the idea of a father choosing a genetic "gift" for his son is repellent . . . it sends all the wrong messages.

—SCIENTIST REVIEWERS OF A PLAY ENTITLED
THE GIFT, WHICH IS BASED ON DILEMMAS
RAISED BY SELECTION FOR TRAITS LIKE
ATHLETIC TALENT

If procreative liberty gives women the right to abort through the first two trimesters for any reason whatsoever, it is hard to see what justification there could be for putting limits on genetic screening and nontransfer of embryos.

—BONNIE STEINBOCK, PROFESSOR OF
PHILOSOPHY, STATE UNIVERSITY OF
NEW YORK, ALBANY

The real problem is not the one we most fear: a government program to breed better babies. The more likely danger is roughly the opposite; it isn't that the government will get involved in reproductive choices, but that it won't. It is when left to the free market that the fruits of genome research are most assuredly rotten.

—DIANE PAUL, PROFESSOR OF POLITICAL
SCIENCE, UNIVERSITY OF MASSACHUSETTS

We mold and shape our children according to environmental factors. We give them piano lessons and every other type of lesson imaginable. I'm not sure there is anything wrong with using genetics . . . as long as it is not hurting anyone or . . . ideas of perfection (are not being imposed) on anybody.

—ARTHUR CAPLAN, DIRECTOR OF THE CENTER
FOR BIOETHICS AT THE UNIVERSITY OF
PENNSYLVANIA

A CAUTIONARY NOTE: GENES ARE ONLY PART OF THE STORY

Imagine the personal disappointment that parents might feel if the trait they've selected in an embryo ends up not being expressed in their child. For disease traits that are genetically determined—like cystic fibrosis or Tay-Sachs—this should never be a problem (as long as technical errors can be avoided); if at least one good copy of the gene is selected, the disease cannot be expressed. But once selection ventures outside the realm of these simple traits, nongenetic factors may play a role as well.

Someday parents will be able to select against alleles that allow a person to become physiologically addicted to alcohol. But even in the absence of such alleles, some people will still drink to excess for purely psychological reasons. And while it's possible right now to select against mutations in the BRCA1 gene that cause a twenty-fold increase in risk of breast cancer, even those without the mutation still have a 3 percent chance of getting the disease by the age of fifty-five. And innate talent in music, math, or athletics will not be enough to turn a child into a skilled musician, mathematician, or athlete, if she chooses to ignore the starting points provided within her genes.

Unless parents understand the limitations inherent in genetic selection, some are bound to be disappointed. Especially in the realm of personality and achievement, the environment will play as critical a role as the genome. It seems incumbent upon those who offer embryo selection services to make sure that prospective parents understand the limitations of the technology before they begin.

If nothing is guaranteed, you may ask, why would anyone want to do it? But the same question can be asked of parents who spend large sums of money on schooling, piano lessons, and private tennis tutors. It is always possible for children to reject the dreams of their parents, no matter what those dreams are based on. And all children—whether selected or not—are subject to the capricious nature of the world within which they live. Children born after selection *against* Tay-Sachs, or *for* perfect musical pitch, can be struck down by the ravages of the modern environment just as easily as those born in the absence of selection. There is only so much that parents can do to protect their children. After that, they can just hope for the best.

NEGATIVE SELECTION VERSUS POSITIVE SELECTION

While some people are opposed to all forms of genetic selection, many seem willing to draw a line between *negative selection* against embryos with disease alleles—and *positive selection*—in favor of embryos that carry desirable alleles. According to this common point of view, it is acceptable to select *against* embryos destined to express Tay-Sachs disease, but unacceptable to select in *favor* of embryos that can develop into children who may start off better than average in some way.

Yet the idea that specific uses of embryo screening will fall neatly into categories of negative and positive selection is specious. No matter what is being screened, there will be embryos that are chosen for implantation and others that are not chosen. Those who wish to draw a line will have to do it based on differences between *genotypes*.

A genotype is simply the combination of alleles that someone or some cell carries at a particular gene. The number of possible genotypes is determined by the number of possible alleles that exist at a gene; if there are two alleles, there will be three different genotypes.

Let's consider the Tay-Sachs gene as an example to illustrate this point in more detail. Its three possible genotypes are normal, disease, and carrier. The normal genotype, carried by most people, contains two functional copies of the gene. The disease genotype contains two defective copies of the gene and causes an unpreventable, horrible death by the age of five. The carrier genotype—with one functional and one defective allele—has no adverse effects on health but allows for the possibility of having an affected child (if the carrier marries another carrier).

Anyone who is willing to accept some form of embryo selection will undoubtedly accept screening against the Tay-Sachs disease genotype. But what about further selection against a Tay-Sachs carrier genotype? Why not provide a child with the psychological well-being that comes with the knowledge that she herself won't have to resort to reprogenetic technology when it comes time to have her own child?

In practical terms, it's hard to accept selection against Tay-Sachs disease but not Tay-Sachs carriers for the following reason: the test for the disease will distinguish all three genotypes. Thus, whether one likes it or not, the choice will have to be made between embryos that are carriers and embryos that have a normal genotype. There are no grounds I can think of for choosing a carrier genotype over a normal one. But, if you agree, it means you are willing to accept selection against something other than a disease genotype.

Some may argue that while the carrier genotype does not itself cause disease, it can lead to disease in a second generation, and should be considered in this light. But the chances that a Tay-Sachs carrier will marry another are less than 4 percent, and advance knowledge of carrier status will certainly allow a person to avoid the birth of an affected child. So the main negative impact of the carrier state is psychological rather than physical.

Let's now consider a case in which it is possible to select against a genotype that causes a 12 percent lifetime risk of breast cancer in favor of one with a hundred-fold reduction in risk. In this case, you are indeed selecting against a disease, so this should be considered a valid case of negative selection. But it's not that simple because the 12 percent breast cancer predisposition genotype is the average one found in the population. In this case, negative selection against cancer risk is equivalent to positive selection in favor of a genotype that provides a relative advantage over other women. On what grounds do you accept or reject this particular type of embryo selection?

If you are willing to accept selection against the normal 12 percent lifetime risk of breast cancer, and the less frequent, but non-disease-causing Tay-Sachs carrier genotype, it follows logically that you should be willing to accept selection for any genotype that provides a child with any reduction in risk to any disease, or any increase in her chances of attaining psychological or physical well-being. It's worth remembering that the embryo selected could have come to term naturally in the absence of selection. If your child could have been born anyway with a reduced risk of disease, or increased psychological well-being, why not make sure of it?

The remaining portion of this chapter will be devoted to this critical question. But what I hope to have convinced you of, here at the outset, is the difficulty inherent in drawing a moral boundary between acceptable and unacceptable uses of the technology. It is for this reason that I will consider embryo selection as a single entity in the ethical discussion that follows.

THE GHOST OF EUGENICS

There are some people who equate the early embryo with a human being that is deserving of the same respect as a child or adult, based on the idea that each human embryo contains a human spirit, deposited within it at the time of fertilization. These people are generally opposed to the destruction of any embryos at any time, whether it is through the normal practice of IVF or in response to embryo selection. A scientific critique of this viewpoint was presented earlier and will not be considered further here. Instead, I will focus on ethical concerns raised by people who are willing to accept the traditional practice of IVF—where embryos are chosen randomly for introduction into a woman's uterus—but are troubled specifically by genetic selection.

Once people reject the notion that an early human embryo is equivalent to a human being, the reasons for opposing embryo selection are varied, but they can all be classified under the rubric of eugenics. *Eugenics.* The word causes people to shudder. But what exactly is eugenics and why is it considered so bad? We must answer these questions before it is possible to continue our discussion.

Unfortunately, answers are not that easy to come by. As the political scientist Diane Paul writes, " 'Eugenics' is a word with nasty connotations but an indeterminate meaning. Indeed, it often reveals more about its user's attitudes than it does about the policies, practices, intentions, or consequences labeled. . . . The superficiality of public debate on eugenics is partly a reflection of these diverse, sometimes contradictory meanings, which result in arguments that often fail to engage."

In its original connotation, eugenics referred to the idea that a society might be able to improve its gene pool by exerting control over the breeding practices of its citizens. In America, early twentieth-century attempts to put this idea into practice brought about the forced sterilization of people deemed genetically inferior because of (supposed) reduced intelligence, minor physical disabilities, or possession of a (supposed) criminal character. And further "protection of the American gene pool" was endeavored by congressional enactment of harsh immigration policies aimed at restricting the influx of people from Eastern and Southern Europe—regions seen as harboring populations (which included all four grandparents of the author of this book) with undesirable genes. Two decades later, Nazi Germany used an even more drastic approach in its attempt to eliminate—

in a single generation—those who carried undesirable genes. In the aftermath of World War II, all of these misguided attempts to practice eugenics were rightly repudiated as discriminatory, murderous, and infringing upon the natural right of human beings to reproductive liberty. *Eugenics* was now clearly a dirty word.

While eugenics was defined originally in terms of a lofty *outcome*—the improvement of a society's gene pool—its contemporary usage has fallen to the level of a *process*. In its new meaning, eugenics is the notion of human beings exerting control over the genes that are transmitted from one generation to the next—irrespective of whether the action itself could have any effect on the gene pool, and irrespective of whether it's society as a whole or an individual family that exerts the control. According to this definition, the practice of embryo screening is clearly eugenics. Since eugenics is horrible, it follows logically that embryo screening is horrible.

Although the fallacy in this logic is transparent, it is remarkable how often it is used by contemporary commentators to criticize reprogenetic technologies. A recent book entitled *The Quest for Perfection: The Drive to Breed Better Human Beings* uses this theme over and over again to castigate one reproductive practice after another. But simply placing a eugenics label on something does not make it wrong. The Nazi eugenics program was wrong not only because it was mass murder, but also because it was an attempt at genocide. The forced sterilizations in America were wrong because they restricted the reproductive liberties of innocent people. And restrictive immigration policies directed against particular regions of. the world are still wrong because they are designed to discriminate directly against particular ethnic groups. Clearly, none of these wrongs can be applied to the voluntary practice of embryo screening by a pair of potential parents.

Once we remove ourselves from the eugenics trap, it becomes possible to consider the ethical concerns that surround embryo screening in the absence of anxiety-producing labels. Again, I want to emphasize my intent to consider only those concerns related to genetic selection rather than the random disposal of embryos during the normal process of IVF. I will start out with five general concerns based on concepts of morality and naturalness. I will move on to concerns about the negative impact that embryo screening could have on society, and conclude with a look at what the future might hold.

IT IS IMMORAL TO CHOOSE ONE CHILD OVER ANOTHER

When embryo selection is equated with choosing children, there is a palpable sense of revulsion. It is not hard to understand this feeling. Often in the past, and in some places still, genetic choice is exercised through infanticide. The particular choice made most often in some Third World countries is boy babies over girl babies, who are suffocated or drowned soon after birth. In other societies, it is infants with physical disabilities that are most often killed.

But the analogy of embryo screening to infanticide is a false one. What embryo screening provides is the ability to select genotypes, not children. Today, parents can use the technology to make sure that their *one* child— whom they had always planned on bringing into the world—is not afflicted with Tay-Sachs.

Even in the future, when it becomes possible to draw computer images based on genetic profiles, embryos will still not be *real* children. Virtual children exist only in one's mind, and the consummation of an actual fertilization event is not even a prerequisite for their creation. Once genetic profiles have been obtained for any man and any woman, it becomes possible to determine the virtual gametes that each might produce. Each combination of a virtual male gamete and a virtual female gamete will produce a virtual child. And each one of the trillions upon trillions of virtual children made possible by virtual intercourse between a single man and woman (who may never have met) could be associated with a computer-generated profile as extensive and detailed as those presented for the virtual Alices at the start of this chapter. At the end of the story, however, only one real Alice emerged. And what her parents chose for her were the alleles that she received from each of them.

IT IS WRONG TO TAMPER WITH THE NATURAL ORDER

This concern is expressed by many who are not particularly religious in the traditional sense. Still, they feel that there is some predetermined goal for the evolution of humankind, and that this goal can only be achieved by the current *random* process through which our genes are transmitted to our children. However, unfettered evolution is never predetermined, and not necessarily associated with progress—it is simply a response to unpredictable environmental changes. If the asteroid that hit our planet 60 million years ago had flown past instead, there would never have been any human beings at all. And whatever the natural order might be, it is not necessarily good. The smallpox virus was part of the natural order until it was forced into extinction by human intervention. I doubt that anyone mourns its demise.

EMBRYO SELECTION FOR ADVANTAGEOUS TRAITS IS A MISUSE OF MEDICINE

The purpose of medicine is to prevent suffering and heal those with disease. Based on this definition, it is clear that embryo selection could be put to uses that lie far outside this scope. But medical doctors have used their knowledge and skills to work in other nonmedical areas such as nontherapeutic cosmetic surgery. If we accept the right of medical doctors to enter into nonmedical business practices, we have to accept their right to develop private programs of embryo selection as well.

One could argue that since the embryo screening technology was developed with the use of government funds, it should only be used for societally approved purposes. But government funds have been used in the development of nearly all forms of modern technology, both medical and nonmedical. This association has never been viewed as a reason for restricting the use of any other technology in private profit-making ventures.

EMBRYO SELECTION TAKES THE NATURAL WONDER AWAY FROM THE BIRTH OF A CHILD

Many prospective parents choose not to learn the sex of their child before birth, even when it is known to their physician through prenatal testing. There is the feeling that this choice allows the moment of birth to be one of parental discovery. If a child's characteristics were pre-determined in many more ways than just sex, many fear that the sense of awe associated with birth would disappear. For some, this may be true. But this is a personal concern that could play a role in whether an individual couple chooses embryo selection for themselves. It can't be used as a rationale to stop others whose feelings are different.

WHETHER INTENTIONAL OR NOT, EMBRYO SELECTION COULD AFFECT THE GENE POOL

If embryo selection were available to all people in the world and there was general acceptance of its use, then the gene pool might indeed be affected very quickly. The first result would be the almost-complete elimination of a whole host of common alleles with lethal consequences such as Tay-Sachs, sickle cell anemia, and cystic fibrosis.

There are some who argue that it would be wrong to eliminate these alleles, or others, because they might provide *a hidden advantage to the gene pool*. This is another version of the "natural order" argument, based here on the idea that even alleles with deleterious effects in isolated individuals exist because they provide some benefit to the species as a whole. Those who make this argument believe that all members of a species somehow function together in genetic terms.

This point of view has no basis in reality. It results from a misunderstanding of what the gene pool is, and why we should, or should not, care about it. The concept of the gene pool was invented as a tool for developing mathematical models by biologists who study populations of animals or plants. It is calculated as the frequencies with which particular alleles at particular genes occur across all of the members of a population that interbreed with each other.

Most healthy individuals are not carriers of the Tay-Sachs or cystic fibrosis alleles, and if given the choice, I doubt if anyone would want to have his or her genome changed to become a carrier. So on what basis can we insist that others receive a genotype that we've rejected? There is none. Genes do not function in human populations (except in a virtual sense imagined by biologists), they function within individuals. And there is no species-wide knowledge or storage of particular alleles for use in future generations.

In fact, there is not even a tendency or rationale for a species to preserve itself at all. At each stage throughout the evolution of our ancestors—from rodentlike mammals to apelike primates to *Australopithecus* to *Homo habilis* to *Homo erectus* and, finally, *Homo sapiens*—small groups of individuals gained genetic advantages that allowed them to survive even as they participated in the death of the species from which they arose! Survival and evolution operate at the level of the individual, not the species.

There are some who are not concerned about abstract concepts like the gene pool and evolution so much as they are worried that the genetic elimination of mental illness (an unlikely possibility) would prevent the birth of future Ernest Hemingways and Edgar Allan Poes. This worry is based on the demonstrated association between manic depression (also known as bipolar affective disorder) and creative genius.

This could indeed be a future loss for society. But once again, how can we insist that others be inflicted with a predisposition to mental disease (one we wouldn't want ourselves) on the chance that a brilliant work of art would emerge? And if particular aberrant mental states are deemed beneficial to society, the use of hallucinogenic or other types of psychoactive drugs that could achieve the same effect—in timed doses—would seem preferable to mutant genes. It is also important to point out that the perceived loss of mad genius from future society is virtual, not real. If the manic depressive Edgar Allan Poe were never born, we wouldn't miss *The Raven*. Likewise, we don't miss all of the additional piano concertos that Mozart would have composed if he hadn't died at the age of thirty-four.

EMBRYO SELECTION WILL BRING ABOUT DISCRIMINATION

With the use of embryo selection, prospective parents will be able to ensure that their children are born without a variety of non-life-threatening disabilities. These will include a wide range of physical impediments, as well as physiological disabilities (such as deafness or blindness) and learning disabilities.

Many people with hereditary disabilities have overcome adversity to live long and fruitful lives. These people are concerned that the widespread acceptance of embryo selection against their disabilities could reinforce the attitude that they are not full-fledged members of society, and not deserving of love and attention.

Of course, disabilities can result from either genetic or environmental factors. And one common environmental cause of disability in the past was the polio virus, which resulted in paralysis, muscular atrophy, and often physical deformity. Inoculation of children with the polio vaccine was not generally seen as discriminatory against those who were already disabled. Why should genetic inoculation against disability be viewed any differently?

One difference could be in the access of society's members to the inoculation. The polio vaccine was provided to all children, regardless of class or socioeconomic status, while embryo selection may only be available to those families who can afford it. The philosopher Philip Kitcher suggests that as a consequence, "the genetic conditions the affluent are concerned to avoid will be far more common among the poor—they will become 'lower-class' diseases, other people's problems. Interest in finding methods of treatment or for providing supportive environments for those born with the diseases may well wane."

This is a serious concern. But it is important to point out that the privileged class already reduces the likelihood of childhood disabilities through their superior ability to control the environment within which a fetus and child develops. People who argue that embryo selection should *not* be used to prevent serious childhood disabilities because it's unfair to those families who are unable to afford the technology should logically want to ban access of the privileged class to environmental advantages provided to their children as well. Political systems based on this premise have not fared well at the end of the twentieth century.

The alternative method for preventing inequality is referred to as "utopian eugenics" by Kitcher and is based on the vision of George Bernard Shaw of a society in which all citizens have free and equal access to the same disease-preventing technologies (and environments). Although discrimination would not be based on class differences in this utopian society, it could still be aggravated by the overall reduction in the number of disabled persons.

It's important to understand the nature of the relationship that might exist between embryo selection and discrimination against the disabled. Embryo selection will not itself be the cause of discrimination, just as the polio vaccine could not be blamed for discrimination against those afflicted with polio. All it could do, perhaps, is change people's attitudes toward those less fortunate than themselves. An enlightened society would not allow this to happen. Is it proper to blame a technology in advance for the projected moral shortcomings of an unenlightened, future society?

EMBRYO SELECTION WILL BE COERCIVE

I distinguished embryo selection from abhorrent eugenic policies of the past with the claim that embryo selection would be freely employed in Western society by prospective parents who were not beholden to the will of the state. As a consequence, the use of the technology would not be associated with any restrictions on reproductive liberty.

There are social science critics who say that this claim is naive. They fear that societal acceptance of embryo selection will lead inevitably to its use in a coercive manner. Coercion can be both subtle and direct. Subtle pressures will exist in the form of societal norms that discourage the birth of children deemed unfit in some way. More direct pressures will come from insurance companies or state regulations that limit health coverage only to children who were embryonically screened for the absence of particular disease and predisposition genotypes.

How coercion of this type is viewed depends on the political sensibilities of the viewer. Civil libertarians tend to see any type of coercion as an infringement on reproductive rights. And liberal libertarians would be strongly opposed to policies that discriminated against those born with avoidable medical conditions.

Communitarians, however, may view the refusal to preselect against such medical conditions as inherently selfish. According to this point of view, such refusal would—by necessity—force society to help the unfortunate children through the expenditure of large amounts of resources and money that would otherwise be available to promote the welfare of many more people.

The communitarian viewpoint is considered shocking to many in America today because, as Diane Paul says, "the notion that individual desires should sometimes be subordinated to a larger social good has itself gone out of fashion, to be replaced by an ethic of radical individualism."

EMBRYO SELECTION COULD HAVE A DRAMATIC LONG-TERM EFFECT ON SOCIETY

Embryo selection is currently used by a tiny fraction of prospective parents to screen for a tiny number of disease genotypes. For the moment, its influence on society is nonexistent. In fact, there are many critics who think that far too much attention is devoted to a biomedical "novelty item" with no relevance as a solution to any of the problems faced by the world. But with each coming year, the power of the technology will expand, and its application will become more efficient. Slowly but surely, embryo selection will be incorporated into American culture, just as other reproductive technologies have been in the past. And sooner or later, people will be forced to consider its impact on the society within which they live.

The nature of that impact will depend as much on the political *status quo* and social norms of the future as they do on the power of the technology itself. In a utopian society of the kind imagined by George Bernard Shaw, all citizens would have access to the technology, all would have the chance to benefit from it, but none would be forced to use it. In this vision of utopia, embryo selection would take an entire society down the same path, wherever it might lead. Unfortunately, if future protocols of embryo selection remain in any way similar to those used now, the technology will remain prohibitively expensive, and utopian access would bankrupt a country.

A different scenario emerges if Americans hold fast to the overriding importance of personal liberty and personal fortune in guiding what individuals are allowed and able to do. The first effects on society will be small. Affluent parents will have children who are less prone to disease, and even more likely to succeed (on average) than they might have been otherwise as a simple consequence of the affluent environment within which they are raised. But with each generation, the fruits of selection will accumulate. When Alice and other members of her selected class come together to select the alleles that they place into their own children, they won't have to worry about the many deleterious alleles that their parents wisely eliminated. Instead, they will be able to focus their attention on accentuating the positive attributes already in place. And in every subsequent generation, selection could become more and more refined.

It is impossible to predict the cumulative outcome of generation upon generation of embryo selection, but some things seem likely. The already-wide gap between the rich and the poor could grow even larger as well-off parents provide their children not only with the best education that money can buy, and the best overall environment that money can buy, but the "best cumulative set of genes" as well. Emotional stability, long-term happiness, inborn talents, increased creativity, and healthy bodies—these could be the starting points chosen for the children of the rich. Obesity, heart disease, hypertension, alcoholism, mental illness, and pre-dispositions to cancer—these will be the diseases left to drift randomly among the families of the underclass.

But before we rush to ban the use of embryo selection by the privileged, we must carefully consider the grounds on which such a ban would be based. Is this future scenario different—in more than degree—from a present in which embryo selection plays no role at all? If it is within the rights of parents to spend $100,000 for an exclusive private school education, why is it not also within their rights to spend the same amount of money to make sure that a child inherits a particular set of their genes? Environment and genes stand side by side. Both contribute to a child's chances for achievement and success in life, although neither guarantees it. If we allow money to buy an advantage in one, the claim for stopping the other is hard to make, especially in a society that gives women the right to abort for any reason at all.

These logical arguments have been tossed aside in some countries like Germany, Norway, Austria, and Switzerland, as well as states like Louisiana, Maine, Minnesota, New Hampshire, and Pennsylvania, where recently passed laws seem to prohibit the use of embryo selection for any purpose whatsoever. In these countries and states, no distinction is made between the prevention of Tay-Sachs disease and selection in favor of so-called positive traits.

But if the short history of surrogacy is any guide, all such attempts to limit this technology will be doomed to failure. Many Tay-Sachs–carrying parents will surely feel that it is their "God-given" right to have access to a technology that allowed earlier couples to have nonafflicted children, and just as surely, there will always be a clinic in some open state or country that will accommodate their wishes. And if the technology is available for this one purpose, it will also be available for others.

It certainly does seem that embryo selection will be with us forever—whether we like it or not—as a powerful tool to be used by more and more parents to choose which of their genes to give to their children. But as we shall now see, the power of this tool pales in comparison to what becomes possible when people gain the ability to choose not only from among their own genes, but from any gene that one can imagine, whether or not it already exists.

18

The Designer Child

And now nothing will be restrained from
them, which they have imagined.
— GENESIS 11:6

WHAT'S THE PURPOSE OF GENETIC
ENGINEERING?

*At the crossroads where reproductive technology and genetic forecasting
meet, stunning developments are taking place. Work is already under
way to alter the DNA of pre-embryos with genetic defects. Using a
glass needle thinner than a human hair, scientists at East Virginia
Medical School in Norfolk withdrew a single cell from a pre-embryo
(a fertilized egg) in a test tube and analyzed its DNA for the deadly
Tay-Sachs disease. After altering the defective gene they implanted the
pre-embryo in the mother's uterus. Result: a healthy baby girl.*

Although this 1994 quote suggests otherwise, genetic engineering of
human embryos has yet to occur (as of July 1997). The journalist who
wrote the article for *Family Circle* got it wrong. But you can't blame her
for the mistake. It certainly appeared as though genetic engineering had

taken place. In actuality, though, the healthy baby girl resulted from embryo selection, not gene alteration.

Embryo selection will almost always provide an extremely effective method for preventing the appearance of serious genetic diseases—the kind that kill or incapacitate people before they reach their prime. The rationale for this supposition is simple: If two people are healthy enough to have reached a stage in their own lives at which they can contemplate becoming parents, then they must have the capacity to produce embryonic genomes that will provide offspring with at least the same potential.

Selection alone will prevent the inheritance of genotypes that cause cystic fibrosis, Tay-Sachs, Huntington disease, sickle cell anemia, PKU, and hundreds of other metabolic diseases—as long as it is possible for parents to produce a nondiseased genotype. This will always be possible whenever one parent is disease-free. And even when both parents have a dominant disease genotype with a mutation at the Huntington disease gene, for example, it is still possible to identify those embryos (25 percent of the total) that have received a normal allele from both parents and will, therefore, be disease-free. Whenever there is a choice, embryo selection will be preferable to genetic engineering, which is both technically and ethically more problematic (for the time being).

It is only when two parents both carry genotypes with two defective alleles of the same gene that embryo selection will not work. The disease genotypes we're talking about here can't be any of those associated with the hundreds of metabolic disorders that always cut life short before adulthood is reached. But a small number of serious disorders have responded to medical treatments that can now extend the life spans of those affected into their third decade. In rare instances, two survivors of cystic fibrosis or sickle cell anemia, for example, may decide to marry and have children together. If so, all of their naturally produced embryos would be alike in carrying the same disease genotype present in both parents. Only genetic engineering could rescue their children.

Much more commonly, two prospective parents will both carry alleles causing milder forms of disease that do not typically prevent people from reaching adulthood or having children. Diabetes, heart disease, obesity, myopia, asthma, a predisposition to some cancers, and many other conditions that adversely affect the functioning of a human organ, tissue, or physiological system are examples. And preemptive cures for all could be achieved by genetic engineering.

The severity of all conceivable diseases extends over a broad range from nearly inconsequential to life-threatening. The symptoms of some—like mild myopia—can be eliminated entirely with proper treatment, such as prescription glasses. The symptoms of others—like mild forms of obesity—may not be life-threatening but can affect a person's quality of life. And others still—like heart disease—are associated with an increasing risk of death as a person grows older.

There are some who believe that it will be possible to draw a moral boundary between acceptable and unacceptable uses of genetic engineering. For most who hold this point of view, curing diseases is considered acceptable while attempts at "genetic enhancement" are not. But as we saw with embryo selection, it is impossible to draw a line in an objective manner. In every instance, genetic engineering will be used to add something to a child's genome that didn't exist in the genomes of either of its parents. Thus, in every case, genetic engineering will be genetic enhancement—whether it's to give children something that other children receive naturally, or to give them something entirely new. It is for this reason that I will use the term "genetic enhancement" interchangeably with "genetic engineering" and "gene therapy."

Genetic enhancement has yet to be attempted on human embryos destined to be children, even though its application to mice and other mammals is now routine. What is it that stops reprogeneticists from hurtling over this final biomedical frontier? Is it technological limitations, ethical concerns, or both? Let's try to answer this question before we move on to the ways in which genetic enhancement might be used in the future, and how it might affect our species.

THE TECHNOLOGY OF GENETIC ENHANCEMENT

Genetic engineering of single-cell bacteria was first accomplished in 1973 and is now a routine tool of the biotech industry. Foreign genes can be added by sprinkling DNA over a dishful of bacteria in a solution that causes little holes to appear and disappear rapidly along the cell surface. If the DNA is in the right place at the right time, it can pass through one of these holes and into the bacterial cytoplasm. This event will typically occur in only one cell out of many thousands. But scientists have devised simple methods for identifying and isolating the infrequent cells that have taken up the foreign DNA. As a consequence, the genetic engineering of bacteria is effective, efficient, and easy to accomplish.

When it comes to multicellular organisms, like mice or humans, the only way to make sure that every cell in a body receives the foreign gene is to transfer it in at the one-cell embryo stage. Sprinkling DNA onto 10,000 embryos with the hope of recovering the one that consumes it is not a practical method of genetic engineering.

An alternative method was developed in 1980 by Professor Frank Ruddle and his student Jon Gordon working together at Yale University. They used a special microscopic needle containing foreign DNA to poke right through the membrane and cytoplasm of the one-cell mouse embryo, and into a pronucleus, where the foreign DNA was released. The added DNA became incorporated into one of the embryo's chromosomes, and was copied faithfully through each cell division into every cell in the adult body, and passed on to the animal's progeny. Ruddle and Gordon's accomplishment demonstrated, for the first time, that the genetic engineering of human embryos was no longer in the realm of science fiction.

Animals that carry foreign genetic material placed by scientists into their genomes (or the genomes of their ancestors) are referred to as "transgenic," and the foreign genes themselves are called "transgenes." Within just a few years after Ruddle and Gordon's report, transgenic technology had spread to laboratories around the world, and the 1986 publication of a lab manual entitled *Manipulation of the Mouse Embryo* provided a step-by-step cookbook-like description of the protocols for all who wished to learn them. By 1997, hundreds of thousands of transgenic mice, pigs, cows, and sheep had been produced by embryo injection with a variety of foreign pieces of DNA.

It might seem as though the currently available transgenic technology should provide a powerful tool for the genetic alteration of human embryos. So why aren't reprogeneticists using it? You might assume that it's ethical enlightenment that stops them. But ethical concerns have done nothing to stop the use of other reprogenetic practices like surrogacy. Instead, the real reason is more mundane. It is because of technical problems that don't bother animal embryologists very much, but would be unacceptable to potential human clients.

First, of all the embryos injected with foreign DNA, less than half actually incorporate the DNA into their chromosomes. If a DNA fragment is not incorporated into a chromosome, it is not copied properly and disappears early during embryonic development. By itself, this low rate of success would be unacceptable to most potential parents, but there's a second problem that's even more severe.

The second problem is that when incorporation into a chromosome does occur, it occurs randomly. This usually doesn't cause any damage because, surprisingly, 95 percent of the DNA in the genomes of all mammals (including humans) doesn't actually serve any purpose for the individual organism. When a transgene inserts itself into these nonfunctional DNA regions, no harm is done. But 5 percent of the cell's DNA is associated with genes, and when a piece of foreign DNA inserts itself into the middle of a gene, it destroys the ability of that gene to function. Thus, 5 percent of all transgenic mice carry a new mutation in a random gene. With humans, a risk factor at this level is unacceptable for any treatment of embryos that are destined to become children.

In the near future, a new method of genetic engineering will be perfected that eliminates the problem of gene disruption. Instead of injecting naked genes into embryonic nuclei, future reprogeneticists will gently put in whole chromosomes that "mind their own business." These "artificial chromosomes" will be constructed in the laboratory with components that ensure their faithful duplication and passage into the pair of cells that forms with every cell division in the developing embryo and fetus. A critical advantage of artificial chromosomes is that they provide a means for adding not just one gene to an embryo, but a "gene-pack" containing hundreds, even thousands, of new genes with many different properties.

There will, however, still be a problem with quality control: It's always possible that the chromosome will be damaged upon injection or that the embryo will reject it. But there's a simple solution and, like so many other technical advances in reprogenetics, it is achieved by bypassing the problem rather than solving it directly. Instead of improving the technique, all one has to do is identify and use just those embryos that have been modified successfully. Reprogeneticists could start out by injecting the artificial chromosome into a dozen or more fertilized eggs, and then allowing them to develop to the eight-cell stage. At this point, one cell from each embryo would be removed for genetic testing to determine whether the new chromosome was present and undamaged. Only properly modified embryos would be chosen for introduction into the mother's womb.

One limitation of the transgenic technology (as it is now practiced) is that it only provides a means for *adding* genes to the genome, and not for altering genes that are already there. In many instances, though, the goal of reprogeneticists will be to eliminate the expression of a wayward gene and replace it with a normal one. An example of this is sickle cell disease, which is caused by the production of mutant hemoglobin proteins inside red blood cells. The most obvious way to cure this disease—in embryos produced with gametes from two sickle cell parents—is to *replace* the gene coding for the mutant proteins with a gene that codes for normal proteins. This goal cannot be accomplished by just injecting a normal gene (because the mutant proteins would still be made).

Gene replacement has already been achieved in animals with a rather different approach from the transgenic protocol described above. For human applications, the entire protocol would work as follows. First, an embryo obtained by IVF would be grown under laboratory conditions into a mass of millions of embryonic-like ES cells. Second, a DNA fragment that contained a replacement version of a particular human gene would be sprinkled onto the ES cells under conditions that allowed fleeting holes to open up in the cell membranes. As a result, the added DNA could enter a cell's cytoplasm and pass through to the nucleus, where it could then kick out and replace the original version of the same gene within the cell's chromosome.

Believe it or not, this entire sequence of events can actually occur. But, as you might imagine, it is exceedingly rare, with success achieved in only one in a million cells exposed to the DNA. Fortunately, reprogeneticists have devised methods for easily detecting that one cell in a million, which is the third step in the process. The properly altered cell can be picked out of the original dish and transferred by itself to a new dish where it would be allowed to grow and divide once again. Finally, individual cells on the new dish could be fused with nuclear-free unfertilized eggs to produce an unlimited number of embryos that all carry the same gene modification. Gene replacement technology has already been used success-fully to produce mice having thousands of specific genetic alterations, and there is no technical reason why it couldn't be applied to human cells as well.

But even as I write these words, experimental reprogeneticists are developing a simpler method of gene replacement technology that works around the problem—once again—rather than addressing it directly. This new approach is called "anti-gene therapy," and it is based on the use of transgenes that act to nullify the action of other specific genes. Based on this approach, an anti–sickle-cell gene and a normal hemoglobin replacement gene could both be added together—as a gene-pack—into an embryo with a sickle cell disease genotype. The anti-gene would prevent the production of sickle cell protein while the normal transgene would make normal protein to take its place. The child that emerged from this embryo would be completely healthy even though he would still carry two defective sickle cell alleles (that are now silenced).

Which approach to genetic engineering will end up as the method of choice for future prospective parents? At this point, the direct placement into single-cell embryos of artificial chromosomes with packs of genes and anti-genes seems most promising. However, based on the history of rapid advances in this field, I wouldn't be surprised if new and improved methods for genetic engineering are devised in the future that completely supersede the methods in practice (with animals) today. No matter what technique, or techniques, are ultimately used, genetic engineering of human embryos is sure to become feasible, safe, and efficient by the middle of the twenty-first century. When that happens, we will come face to face with the ultimate frontier in medicine and philosophy—the power to change the nature of humankind.

TREADING IN GOD'S DOMAIN

Genetic engineering has been attacked on many of the same grounds used to attack embryo selection. We hear that it's a dangerous idea with "eugenic potential," and that its use will be an assault on the freedom and dignity of human beings. We're warned that it could harm the gene pool, and impose choices of prior generations on distant offspring. We're informed that it's an improper use of medicine, and an unfair drain on societal resources. We are told that it will discriminate against the disabled and be unfair to those who can't afford it. And we're advised that it will only be used by heartless parents who will treat their children like commodities to be purchased and used.

As is so often the case with new reproductive technologies, the real objection lies in the realm of spirituality, not science. Simply stated, there is a commonly held sense that genetic engineering crosses the line into

God's domain. And as we have all been taught, it is wrong to tread in God's domain.

Throughout history and in nearly every culture, cautionary tales have emerged with the same theme. Adam and Eve ate the forbidden fruit and were banished from the Garden of Eden; the builders of the Tower of Babel came too close to heaven and suddenly spoke in different tongues; Prometheus stole fire from Zeus (the supreme Greek god) to give to man and was chained to a rock for the rest of his life with his immortal liver eaten daily by an eagle; Pandora's curiosity unleashed all the evils from her opened box onto humankind; and Dr. Frankenstein died at the hands of his human creation. Time and again, we are warned of places we should not go, and things we should not do. And while the names may change, the message is still the same. Today, many in the modern secular world believe it is wrong to mess with "Mother Nature," an updated feminine personification of God himself.

As one might expect from the inability of human beings to reach agreement on the meaning of God and his relation to humankind, the line between the domain of man and the domain of God has been drawn at different places by different people. Today, the most expansive view is promoted by groups like the Christian Scientists, who reject all forms of medical treatment for disease. The entire human body is within God's domain, they feel, and not to be touched by mortal men of medicine.

A less expansive view is espoused by the Catholic church, which generally accepts the use of medicine to cure disease but rejects all forms of noncoital reproduction, as well as birth control. So while the human body may lie on the outside of God's domain, the entire process of reproduction is on the inside.

For most people in modern Western society, God's domain has been reduced to a much smaller size, owing in large part to knowledge and use of both birth control and currently available reprogenetic technologies. By 1994, fully 75 percent of all Americans accepted the use of IVF as a treatment for infertility. This sets a boundary that extends no further out than the surface of the fertilized egg itself. And for the most part, those who approve of IVF also accept the use of ancillary technologies, like the injection of testis-derived nuclei into the egg cytoplasm. The acceptance of this last practice, in particular, reduces the domain of God even further down to the surface of the DNA-containing nuclei floating serenely within the egg cytoplasm.

You can see the problem we are running into. If we allow the possibility that "man's domain" extends into the nucleus—into the DNA itself—then by this line of reasoning, God's domain vanishes into . . . nothingness. This frightening notion compels some people to draw a final line—a veritable last stand—around the genetic material. Indeed, 45 percent of Americans reject all uses of genetic engineering even when it is needed to cure a serious disease, and 85 percent reject its use for any purpose other than a disease cure. The British public is even more wary of the technology with 89 percent rejecting its use "to improve intelligence," and 95 percent rejecting its use to achieve "good looks."

While almost half of all Americans reject the use of genetic engineering to cure disease, it goes without saying that nearly all (except Christian Scientists) would accept the use of medical treatment—not directed at genes—to obtain the same result. And four to five times as many British respondents thought it would be appropriate to use something like "vitamins," rather than genetic engineering, to increase intelligence or good looks. Of course, nearly everyone would accept the use of orthodontics to straighten teeth, rhinoplasty to straighten noses, and good nutrition and education to enhance intelligence.

Although all other intrusions into the body may work around the edges, genetic engineering, it seems, impinges on the essence of life itself—the soul. And the soul is clearly in God's domain. The sociologists Dorothy Nelkin and Susan Lindee describe it well:

> "Just as the Christian soul has provided an archetypal concept through which to understand the person and the continuity of self, so DNA appears in popular culture as a soul-like entity, a holy and immortal relic, a forbidden territory. The similarity between the powers of DNA and those of the Christian soul, we suggest, is more than linguistic or metaphorical. DNA has taken on the social and cultural functions of the soul. It is the essential entity—the location of the true self—in the narratives of biological determinism."

There is a serious flaw in the apparently logical progression that leads people down, step by step, to the idea that the *essence* of human life is contained within the genetic material. The flaw is caused by the inability to separate the two very different meanings of the word *life* described at the beginning of this book—at the level of the individual cell and at the level of consciousness. DNA may very well represent the *essence* of a cell's life. But human life—in the special meaning of the term—does not exist in the single-cell embryo or in any single neuron. Human life emerges only at a higher level, when the trillions of cells in the brain all function together. The essence of human life lies within the human mind, not within inert molecules of DNA. Whether the human mind should be viewed as part of God's domain is, for the time being, a question of faith, not science.

Although human essence does not lie within DNA, some future applications of genetic engineering could powerfully affect the essence of those people who emerge with modified genes. We know this is true from the effects that genetic modifications of the past have had. Five million years ago, embryos indistinguishable from those that gave rise to you and me, with genomes 99 percent the same as ours, produced hairy apes that had no human essence at all. A genetic modification of just 1 percent was all it took to create minds with the ability to contemplate their own consciousness, minds with the ability to contemplate further genetic modifications that could enhance the minds of future human beings.

Human essence came into existence simply because those with it could out-compete and kill those without it. But if human minds have the ability to contemplate and direct changes in the copies of their own genomes that they give to future generations, the human mind is much more than the genes that brought it into existence. While selfish genes do, indeed, control all other forms of life, master and slave have switched positions in human beings, who now have the power not only to control but to create new genes for themselves.

Why not seize this power? Why not control what has been left to chance in the past? Indeed, we control all other aspects of our children's lives and identities through powerful social and environmental influences and, in some cases, with the use of powerful drugs like Ritalin or Prozac. On what basis can we reject positive genetic influences on a person's essence when we accept the rights of parents to benefit their children in every other way?

THE FUTURE

Genetic engineering will eventually be used by future reprogeneticists. It will begin in a way that is most ethically acceptable to the largest portion of society, with the treatment of only those childhood diseases—like sickle cell anemia or cystic fibrosis—that have a severe impact on quality of life. The number of parents who will desire this service will be tiny, but their experience will help to ease society's trepidation.

As the fear begins to subside, reprogeneticists will expand their services to nullify mutations that have a less severe impact on a child, or an impact delayed until adulthood. Predispositions to obesity, diabetes, heart disease, asthma, and various forms of cancer all fall into this category. And as the technology spreads, its range will be extended to the addition of new genes that serve as genetic inoculations against various infectious agents, including the HIV virus that causes AIDS. At the same time, other genes will be added to improve various health characteristics and disease resistance in children who would not otherwise have been born with any particular problem.

The final frontier will be the mind and the senses. Alcohol addiction will be eliminated, along with tendencies toward mental disease and antisocial behavior like extreme aggression. Visual and auditory acuity will be enhanced in some to improve artistic potential. And when our understanding of the genetic input into brain development has advanced, reprogeneticists will provide parents with the option of enhancing various cognitive attributes as well.

Is there a limit to what can be accomplished with genetic enhancements? Are there certain attributes that we will never be able to incorporate into human descendants? Perhaps. There are many experts in genetics and reproduction today who use the word *impossible* to assert limits to future knowledge and technology in this area. But, as the physicist and futurist Freeman Dyson says in this regard, "The human species has a deeply ingrained tendency to prove the experts wrong."

One way to identify types of human enhancements that lie in the realm of possibility—no matter how outlandish they may seem today—is through their existence in other living creatures. If something has evolved elsewhere, then it is *possible* for us to determine its genetic basis and transfer it into the human genome. Relatively simple animal attributes that fall into this category include the ability to see into the ultraviolet range or the infrared range—which would greatly enhance a person's night vision. Other possibilities include light-emitting organs (from fireflies and fish), generators of electricity (from eels), and magnetic detection systems (from birds). More sophisticated animal attributes include the ability to distinguish and interpret thousands of different airborne molecules present at incredibly low levels (through the enhanced sense of smell available to dogs and other mammals), and the ability to generate and sense reflected high frequency sound waves to "see" objects in complete darkness through a biological sonar system (by bats).

Another possible sensory enhancement is four-color vision. Normal people are able to see three colors—red, blue, and green—but some people (and most animal species) are born color blind, with the ability to see only two colors, or just one. People with one-color vision see the world as if they are watching a black-and-white television, perhaps tinted in one color or another. In one version of a two-color world, some color-blind people see only shades of blue and red, without any greens or yellows. Now imagine what would happen if a person who was color blind from birth was suddenly able to see all three colors. It would probably be akin to an hallucinogenic experience. That's the only way to imagine what a four-color world might look like from a three-color perspective.

And then there is *radiotelepathy*, the term used and defined by Freeman Dyson to describe the ability of a person, or creature, to send and receive information as radio waves. Radio waves and visible light are both forms of radiation occupying different zones within the electromagnetic spectrum; the only distinction between the two is the range of energies carried within the individual photons of which each is composed. What our eyes detect as differences in color are simply photons with slightly different energies, which correspond to different frequencies. There is no inherent biological barrier to the development of a sensory organ that could distinguish different radio frequencies instead of light frequencies. And it seems reasonable to assume that a neurological structure could be developed that was dedicated to the interpretation of information in the form of radio amplitude and frequency modulations—the basis for AM and FM radio broadcasts, respectively—just as the auditory cortex within our own brains allows us to interpret modulations in sound—which we *hear* as language and music. More problematic, perhaps, is the development of a biological organ that could *emit* modulated radio waves. But even this could be imagined as a sophisticated enhancement of the light-producing systems used by fireflies and deep water fish.

In the short term, though, most genetic enhancements will surely be much more mundane. They will provide little fixes to all of the naturally occurring genetic defects that shorten the lives of so many people. They will enrich physical and cognitive attributes in small ways. And as the years go by over the next two centuries, the number and variety of possible genetic extensions to the basic human genome will rise exponentially— like the additions to computer operating systems that occurred during the 1980s and 90s. Extensions that were once unimaginable will become indispensable . . . to those parents who are able to afford them.

Epilogue
Human Destiny?

> I am Alpha and Omega, the beginning and
> the end, the first and the last.
> —REVELATION 22:13

DATELINE WASHINGTON, D.C.: MAY 15, 2350

The commission of leading academics—established by Dr. Albert Varship six months earlier—had come to Washington, in secrecy, to present their final report. One representative from each of the relevant fields—the reprogeneticist, the evolutionary biologist, the demographer, the sociologist, and the psychologist—sat around the table in the conference room at the Department of Health and Human Services. One by one, they took turns presenting a portion of the report to the HHS Secretary.

Their findings were grim; their predictions were surreal. Yet, Dr. Varship could find no flaw in their logic, no reason to challenge the central conclusion in their final joint summary statement: "If the accumulation of genetic knowledge and advances in genetic enhancement technology continue at the present rate, then by the end of the third millennium, the GenRich class and the Natural class will become the GenRich-humans and the Natural-humans—entirely separate species with no ability to cross-breed, and with as much romantic interest in each other as a current human would have for a chimpanzee."

The presentation took just over two hours. Throughout, Dr. Varship sat in silence. It was too horrific to comprehend. Unbelievable, and yet, entirely predictable. Indeed, predicted long, long ago.

Dr. Varship's mind wandered back to his teenage years, when he had been an avid reader of science fiction, including stories written by one of the fathers of the field—H. G. Wells—at the end of the nineteenth century. So much of what Wells had prophesied—television, intercontinental air travel, space stations, motion pictures, air-conditioned cities, and much more—had become real early on. And now this as well—"the splitting of the human species." Wells had written, "the gradual widening of the present merely temporary and social difference between the Capitalist and the Laborer was the key to the whole position," in the antiquated political language of that era. Now it was all coming true.

The only thing that Wells got wrong was how long it would take. Space travel to other worlds was one thing, but the notion that humans might someday be able to manipulate their own genes was clearly too ludicrous to consider during the first half of the twentieth century, even by visionaries like Wells, Verne, Huxley, and Asimov. And yet here we were on the cusp of an incredible evolutionary event. Not in the way Wells had imagined—as the result of natural evolution, 800,000 years hence—but in less than a millennium as a result of self-evolution.

It had been three hundred years since genetic enhancement began in earnest. During that time, twelve generations of GenRich individuals had lived and reproduced. With each generation, it became possible to start with an already-enhanced genome that could be enhanced even further. And with each generation, an increase in biomedical understanding and genetic technology allowed reprogeneticists to make ever more complex enhancements, with hundreds, sometimes thousands, of added genes.

Although the initial focus was on physical and mental health, it shifted quickly to personality traits and talents in the cognitive, athletic, and artistic realms. In these areas, different enhancements were chosen for different GenRich children. But these differences sat on top of an ever-expanding genetically enhanced framework that was shared by all members of the GenRich class.

Varship was frightened by what he heard, and searched for the right response. Genetic enhancement clinics—GE centers, as they were popularly known—were spread across North America. They were all run as private enterprises without any government assistance. Indeed, long-existing laws prohibited the use of federal funds for what was euphemistically called "research" on human embryos. Elected officials and GE executives both found this prohibition convenient for political cover, and it provided the basis for the "hands-off" approach that the government had consistently taken toward GE. It was for this reason that Varship had formed his commission in secrecy. But now that their final report was in his hands, what could he do with it?

The problem was that GE represented a multi-billion-dollar industry that served not only American citizens, but many foreigners as well. Indeed, the American GE industry benefited enormously from restrictive laws that limited its practice in many other countries, and as a consequence, this single industry had a major impact on reducing the balance of trade on the side favorable to the American economy. Not surprisingly, politicians and their supporters from the business community were loath to go anywhere near it. Of course, over the years, common citizens had occasionally expressed their concern about the long-term societal impact of GE. Rights to privacy; individual liberties; the folly of governmental intrusion into the free market—these were the talking points that politicians focused on in response to such concerns.

Varship and all of the presenters in the room with him that morning were themselves GenRich. If they had been born otherwise, they would never have attained the positions they held. All members of Congress, all entrepreneurs, all other professionals, all atheletes, all artists, and all entertainers were members of the GenRich class. There was no longer any way that even the most talented Natural could advance into any of these realms.

What could be done? What was possible? Put a stop to the whole thing, there and then? Outlaw the practice of Genetic Enhancement? There would be an outcry from all the GenRich. A Congress filled with GenRich legislators would never allow it to happen. And even if it did come to pass, in the end, it would make no difference. Sure, it might slow things down in the short term—perhaps a few months—but GE centers would simply move to off-shore islands, and to underdeveloped countries eager for added tax revenue. The prospective GenRich parents would all follow them abroad.

If legal restrictions erected in one country or another were useless, was there another way to stop the practice of GE? Varship considered the moral argument. Perhaps he could convince the President—who underneath his tough political skin showed twinges of humanity—to bring his enormous influence to bear on the problem and preach the sins of GE. Perhaps a campaign could be undertaken to explain to all GenRich people the frightening moral consequences of GE for humanity as a whole.

Unconsciously, Varship shook his head as he realized the elimination of GE was hopeless. All prospective parents wanted to provide their children with the greatest possible advantages in life. It had been that way for hundreds of thousands of years. How could you convince parents to forsake this instinctive personal desire for the good of society? Each individual parent would say, "The genetic enhancement of just my child has no impact on society at all. Why is it immoral for me to want the best for my children? I'm not harming anyone else by my actions."

So much had changed, and so much would have to change again to get back to the way things once were (if ever they really were so). The gap between the GenRich and Naturals lay not just in genes, but in every other aspect of their lives and communities and, most important, in their monetary resources. Stopping the practice of GE cold, at this point in history, would not bring the classes back together again.

If there was no way that GE could be halted, was there a way to stop it from breaking humankind into two? Varship imagined a utopian society in which GE was freely available to all, and where all Naturals were raised to the level of the GenRich. It brought a moment's smile to his face, but just a moment and not more. Santa Claus existed only in the minds of children, and there was no way a society could afford to provide this expensive service to all of its citizens, even if it wanted to.

Where had we gone wrong? Was there any time in the past when a different course might have been pursued? Varship was well-versed in the early history of GE. The original practitioners drew a moral line between preventing disease and enhancing characteristics. How could anyone argue against preventing childhood disease? But it soon became clear that the moral line was an imaginary one. It was all genetic enhancement. It was all done to provide a child with an advantage of one kind or another that she would not have had otherwise. And what was wrong with that? What was wrong with helping children to live better lives?

The history books made it clear that early twenty-first-century scientists had failed to see the cumulative impact of GE. Even as scientific understanding and technology continued to explode exponentially around them, they continued to assume that the future would be the same as the present, and that complex physical and cognitive traits would always be beyond reach. With a shock that opened his eyes wide, Varship realized that most present-day scientists had the same mental block as their predecessors.

It was late, by Varship's reckoning. Too late to do anything at all, he concluded helplessly. We were on a journey into a rapidly evolving future that no man, no woman, could stop. And where it might lead, no one could tell.

DATELINE THE MILKY WAY: JUNE 1, 2997

Just as Dr. Varship had suspected 647 years earlier, his scientific colleagues had been woefully conservative in their predictions of where GE would lead humankind. It was all because they had failed to appreciate the power of exponential advancement—not just in technology but in the essence of the human species itself.

Even simple cumulative processes had a way of taking early scientists by surprise. By the end of the nineteenth century, evolutionary biologists knew that their species could trace its ancestry back along a direct line to an apelike mother who had lived 5 million years earlier, and whose children had gone on to generate both human beings and chimpanzees. Nowhere along those lines of a million generations did any child appear to be very different from its parents. And yet at the beginning there was an ape; at the end of one line, there was a human being.

Spectacular changes occurred even more rapidly when early humans consciously intervened in the cumulative process. Within a hundred generations, they took individuals from a single species of gray wolves and bred them down different pathways into French poodles and Saint Bernards, into hounds and sheep herders, and into so many other breeds that look and behave so differently it's hard to believe they are all distant cousins of one another.

All of this was known by the end of the twentieth century. Furthermore, significant progress had already been made, at that time, in the major scientific areas that together formed the basis for GE. Scientists were well on their way toward an understanding of how each gene in the human genome functioned. Genetic engineering had already been accomplished with other mammalian species. A prototype of the artificial human chromosome had already been invented. Surely, those who watched these advances take place must have realized where it would all lead. How could the biologists themselves be so blind as to not understand that changes in their own species—predetermined at every step—would occur even more rapidly than the random changes imposed on domesticated animals and plants by earlier people?

But instead, conservative naysaying scientists ruled the day. Yes, the biologists admitted, we will soon identify every human gene. But, we'll *never* truly understand how all these genes interact with one another during the development of a human life. Although the human genome provides a blueprint, the blueprint is indirect and impossible to read in any context other than the developing human embryo and fetus. This is because each one of the billions of cells in the fetus acts as its own little computer in interpreting the genetic instructions present in its DNA in the context of its own little microenvironment. As a consequence, the biologists said, it would be *impossible* for even the most powerful computer to simulate the development of a human being starting with just the information present in a one-cell embryo. And because of this, they went on, big changes to the human genome would never be attempted since reprogeneticists would have no way of knowing *ahead of time* how these changes would really affect the child who was born with them.

But these late twentieth-century scientists made the same mistake as so many of their predecessors. Understanding the true nature of the gene is "beyond the capabilities of mortal man," they said in 1935; it is *impossible* to determine the sequence of the complete human genome, they said in 1974; it is *impossible* to alter specific genes within the embryo, they said in 1984; it is *impossible* to read the genetic information present in single embryonic cells, they said in 1985; it is *impossible* to clone people from adult cells, they said in 1996. And all of these impossibilities not only became possible but were accomplished while the early naysayers were still alive.

It is hard to believe they couldn't see that not only would all genetic interactions be uncovered, but that computers *would* become powerful enough to simulate the effects of any imagined genetic alteration or addition to the genome. (Now, of course, no GE engineer would ever dream of adding a new gene-pack to an embryo without first testing its effects by computer simulation.)

In the twenty-fourth century, Dr. Varship's commission predicted that humans would diverge into just two species—the GenRich and the Naturals. Naturals had the standard set of 46 chromosomes that long defined the human species, while the GenRich alive at that time had an extra pair specially designed to receive additional gene-packs at each new generation. With 48 chromosomes and thousands of additional genes, the GenRich were, indeed, on their way to diverging apart from the Naturals.

But what twenty-fourth-century reprogeneticists failed to see was the looming consolidation and competition within the GE industry, and the impact of the earth's population explosion. Until the end of the twenty-fourth century, reprogeneticists had agreed to use the same special chromosome as the platform for all their enhancements. But in the twenty-fifth century, everything changed. Independent GE centers around the world were bought up by one of the three giants—Microgene, Unigene, and Macingene. Soon thereafter, in the heat of intense competition, each corporation began to modify the chromosomes offered to their clients in different, incompatible ways. As a result, GenRich families enhanced at Microgene-owned clinics began to diverge from those enhanced at Macingene-owned clinics, and both began to diverge from those enhanced at Unigene-owned clinics. By the twenty-sixth century, the original species of *homo sapiens* had already evolved into four separate species, not two. And that was just the beginning.

In the twenty-sixth century, overcrowding on earth had reduced the quality of life so much that many GenRich parents decided to give their children special genetic gifts to help them survive on worlds that were inhospitable to the unenhanced. The development of these new gene-packs was based partially on genetic information obtained from various creatures living under extreme conditions on earth—including giant clams, tube worms, and microscopic bacteria that thrive in scalding hot sulfurous water around volcanic vents on the ocean floor, far removed from light and free oxygen; and other creatures that use a biological form of antifreeze to thrive around Antarctica. In addition, GE engineers had achieved human symbiosis with plants through the successful incorporation of photosynthetic units into embryos. Not only could symbiotic humans receive energy directly from the sun, but they were now able to self-produce some of their own oxygen from water and carbon dioxide, just like plants.

The new era of exploration began with settlements at the edge of the ice-covered northern polar cap of Mars. The lung-modified thick-skinned dark green human descendants that began their lives on the fourth planet from the sun barely resembled the primitive Naturals still roaming the third planet Earth. Of course, these green people had made sure to arrive with a variety of specially engineered animal-plant creatures that were also uniquely adapted to their new world. Some were used for food, others as pets, and others still were designed to extract large quantities of oxygen from the frozen water (using sunlight as an energy source) for the maintenance of optimal living conditions within enormous bubble-enclosed biospheres.

As Earth's population continued to expand, other types of enhanced GenRich groups moved to other planets, moons, and asteroids in the original solar system, where they used GE to further enhance the ability of their own children to survive on their chosen worlds. As the first artificial chromosome pair reached capacity, additional chromosome pairs of different types were added into subsequent generations. By the middle of the twenty-seventh century, there were at least a dozen different species of human descendants having chromosome numbers that varied from forty-six in Naturals to fifty-four in the most enhanced GenRich individuals.

It was a long-sought-after genetic enhancement—finally perfected in the twenty-seventh century—that made it possible to even *think* about traveling to other solar systems. This was the gene-pack—designed by Macingene—that slowed the aging process down to a crawl. Children born with the AGEBUSTER gene-pack would live for hundreds of years, perhaps longer, with minds and bodies intact. Like young explorers throughout all the centuries of human existence before the twentieth, they said good-bye to their families knowing they would never see them again, and boarded enormous citylike nuclear-powered spaceships to travel to inviting planets discovered by astronomers in nearby solar systems.

And now here we sit in the year 2997 and ponder the future. Ongoing enhancements in the AGEBUSTER gene-pack and the technology of space travel are certain to expand the reach of human life across our galaxy, and perhaps beyond. With this expansion, far-flung communities will begin to lose contact with one another. Indeed, many will lose cultural memory of their species origin on the third planet in a nondescript solar system lost among the billions in the Milky Way. Eventually, the descendants of humankind will travel through millions of centuries, explore millions of worlds, and diverge into millions of different species with little resemblance to the humans of the twentieth century . . . as they recapitulate the many paths followed by that very first cell—the mother of *all* living things—on the planet Earth, so long ago.

DATELINE THE UNIVERSE: ????

The most incredible thing about the original human genome was that it provided human beings with a human consciousness able to imagine all of the things described in this book. The journey from three billion bases of genetic information to human consciousness was long and contorted, but no scientist from the time of Watson and Crick onward could sincerely doubt that the journey was indeed made during the development of each human being.

The second most incredible thing about the human genome was how readily it revealed its secrets to humankind. The biggest secret, of course, was the precise genetic pathway that led to consciousness and intelligence. There were those who thought that "intelligence genes" would be found by looking for differences in the genomes of so-called smart and dumb people, but this approach was hindered by strong interference from environmental influences. Others thought that answers would come only from a deep understanding of how the brain was wired. But twenty-first-century neuroscientists had neither the tools, nor the mental abilities, to map out or comprehend the trillions and trillions of connections that existed among neurons. In the end, the breakthrough came from an entirely different direction—through a look at our own evolution.

To appreciate the evolutionary approach, it is useful to consider the way in which geneticists generally discovered the root causes of things. The genetic basis for sickle cell anemia was not determined by looking at diseased people alone, it was uncovered by searching for the difference between a diseased person and a healthy one. And in the same way, the genetic basis for human consciousness and intelligence was not discovered by comparing humans to one another, but by comparing the shared human genome to that of their nearest living relative—the chimpanzee.

Incredibly, to the scientists who first took a look, the genome of the chimpanzee was virtually identical to the genome of a human being. In retrospect, this shouldn't have been surprising since the two species were only five million years apart. Yet the human genome gave rise to human consciousness, while the chimp genome gave rise to a primitive form of subhuman consciousness. Clearly, the genetic basis for the greatly enhanced consciousness and intelligence of human beings had to be found among the small number of significant differences that existed between the two genomes.

By the end of the twenty-second century, all of the genetic enhancements that were required (in theory) to provide a chimp with a human mind had been identified, even though it took longer to really understand what a human mind was all about. In the view of some, it was God's gene-pack that had been uncovered. But to reprogeneticists, it was just a marvelous tool. For if a chimp's brain could be converted into a human's "on paper," then further enhancements of *those very same genes* could convert a human brain into something that was that much more advanced.

It was a critical turning point in the evolution of life in the universe. For when the first generation of cognition-enhanced GenRich matured, they produced among themselves scientists who greatly outshone geniuses from all previous epochs. And these scientists made huge advances in further understanding the human mind, and they created more sophisticated reprogenetic technologies, which they then used to enhance cognition even further in the GenRich of the next generation. In each generation hence, there were quantum leaps of this kind. Throughout it all, there were those who said we couldn't go any further, that there were limits to mental capacity and technological advances. But those prophesied limits were swept aside, one after another, as intelligence, knowledge, and technological power continued to rise.

A special point has now been reached in the distant future. And in this era, there exists a special group of mental beings. Although these beings can trace their ancestry back directly to *homo sapiens*, they are as different from humans as humans are from the primitive worms with tiny brains that first crawled along the earth's surface. It took 600 million years for those worms to evolve into human beings. It has taken far less time for humans to self-evolve into the mental beings that now exist.

It is difficult to find the words to describe the enhanced attributes of these special people. "Intelligence" does not do justice to their cognitive abilities. "Knowledge" does not explain the depth of their understanding of both the universe and their own consciousness. "Power" is not strong enough to describe the control they have over technologies that can be used to shape the universe in which they live.

These beings have dedicated their long lives to answering three deceptively simple questions that have been asked in every self-conscious generation of the past.

"Where did the universe come from?"

"Why is there something rather than nothing?"

"What is the meaning of conscious existence?"

Now, as the answers are upon them, they find themselves coming face to face with their creator. What do they see? Is it something that twentieth-century humans can't possibly fathom in their wildest imaginations? Or is it simply their *own* image in the mirror, as they reflect themselves back to the beginning of time . . . ?

Acknowledgments

It is common for authors to pay tribute to the persons *most* important to their writing endeavor at the end of the acknowledgments, much in the way a beauty pageant progresses slowly through many finalists to a single winner. My intention here is to recognize those who were most important in making this book possible at the outset, not the sunset.

This book simply could not have been written without my wife, Susan, and the constant love, companionship, and broad-ranging discourse that has infused our lives together since September 26, 1984. She has been, and always will be, my most trusted critic, the one who tells me how things really are viewed outside the cloistered worlds of science and the academy. If I have succeeded in accomplishing the task I set for myself of informing nonscientists about the nature of science, it is because of her.

My children have provided me with an appreciation for the meaning of life, and much more. Together they have made our family one that exists and persists in the midst of a constant whirlwind of excitement and adventure. While crossing the Sahara desert on tracks that vanish under the sand, sleeping in an underground Star Wars pit, dodging bombs in Corsica, hanging out at a separatist Basque café in Spain, exploring abandoned medieval castles on the Côte d'Azur, traipsing across a Siberian village, climbing a precipice in Maine, or simply dining at an elegant restaurant in Provence or Tuscany, they have reveled in the joy of life and made my own life that much more fulfilling. In addition, I thank my daughter for her dramatic creativity, and never failing to point out the contradictions in my thoughts and actions; my middle son for his exuberance, good nature, and ability to convince me to do things I would not otherwise do; and my youngest for providing a mirror into my long-ago self with his broad-eyed wonder about the way the world is put together.

My parents have always been there to provide support and encouragement, as well as a sense of identity. My brother and sister, Bruce and Susan, as well as brothers- and sisters-in-law Dave, Dave, Jay, Jane, and Lee, my mother- and father-in-law, nieces, and nephews come together as an extended family that is also responsible for providing a stable environment conducive to productive writing.

After my family, the person who played the largest role in making this book happen is my incredible agent, Theresa Park. Theresa took on many roles—cheerleader, adviser, editor, and protector—as she paved the way for this book to be published. My assistant, Barbara Smith, deserves special recognition for building order into a life on the verge of chaos. Barbara always made sure I was where I was supposed to be, and she always knew what I needed before I even asked (if I actually remembered to ask).

I thank my editor at Avon, Rachel Klayman, for providing insightful comments that forced me to crystallize my thoughts and focus my ideas in critical places throughout the book, as well as for the numerous ways in which she shepherded this book to publication, and my editor Dick Marek, for pointing out tics in my prose that I never knew existed.

Certain colleagues, family members, and friends gave the gift of time, taken from their own busy lives, to read and critique portions of this book. In this regard, I recognize and thank Vincenne Adams, Angela Creager, Gideon Rosen, Norman Fost, Basil Remis, Harold Shapiro, and Sherrill Cohen, as well as the Princeton chapter of "Reality Check"— Vincenne, Gideon, and Angela, and Norton Weiss, Hope Hollecher, Charlie Gross, Rena Lederman, Emily Martin, Alison Jolly, and Ben Heller—who critiqued the Virtual Child chapter, and generally opened my mind to the social context within which biomedical science operates.

Other friends deserve special recognition for conversations that helped me shape the scientific, political, historical, and philosophical views presented in this book. I thank Sherrill for her feminist libertarian perspective on reproductive rights; Sheldon Garon for his communitarian counterpoint and deep understanding of politics and societies (large and small); Angela and Gideon for their broad historical and philosophical insight; Madelaine Shellaby and Rich Shapiro for insight into questions of spirit and politics; Jonathon Weiner for sharing so many ideas about evolution and behavior; Gina Kolata for her broad knowledge of biology and medicine and the players involved in its practice today; and Ed Witten and Chiara Nappi for discussions about science, life, and the Princeton school board.

My foray into the realm of science and society began when I audited a course taught by Frank von Hippel at the Woodrow Wilson School at Princeton University. I thank Frank for his early encouragement in this direction. My participation as a member of the Task Force on New Reproductive Technologies established by the New Jersey Bioethics Commission during the 1980s was also a great learning experience. I thank all of the members of the Task Force, but especially Alan Weisbard, Ruth Macklin, Adrienne Asch, and Mary Sue Henifin for their insightful perspectives on issues surrounding surrogacy.

A number of individuals provided me with particular information or ideas that were directly incorporated into this book. Brigid Hogan provided me with access to material accumulated by the Human Embryo Research Panel commissioned by the NIH, as well a wonderful personal anecdote. Alan Trounson confirmed my suspicion that pronuclei did *not* fuse in the one-cell human embryo. John Robertson provided insight into the legal basis for reproductive rights. Sydney Brenner got me thinking about evolutionary genetic approaches to the understanding of consciousness. And Sara Storey of Ridgewood, New Jersey sparked my realization of the confusion that arises when identical twins donate reproductive ingredients to each other.

Then there are the ever-stimulating undergraduate students at Princeton University who have forced me to think about all sorts of things that I hadn't thought about previously. Along these lines, I thank those students who took part in the three small seminar classes I have taught at Princeton: my 1995 freshman seminar "Sex, Babies, Genes, and Choices," my 1991 senior seminar "Genetics and Politics in a Brave New World," and my 1987 Policy Task Force on "Human Reproduction" at the Woodrow Wilson School of Public Affairs and International Policy. I also want to thank the many wonderful undergraduates, graduate students, postdoctoral fellows, and assorted visitors who have passed through my lab in the Department of Molecular Biology.

Many of the ideas presented in this book emerged from years of free-flowing discussions with numerous friends, colleagues, and students not mentioned here, as well as a synthesis of material previously published by other authors. I have attempted to recognize all those who made important contributions to my writing in a general way here, and in a specific way in the endnotes. Unfortunately, memories fade with time, and the precise origin of many ideas is now a mystery to me. I apologize to all those whom I have not properly recognized. In no way can I pretend to claim that the ideas expressed here are solely of my own creation. On the other hand, if there are mistakes in some of the scientific concepts or techniques that I present, for these I must take full blame. I have made an attempt to be as accurate and honest as possible in all discussions of science—present and future—but there are sure to be places where I am just plain wrong. I apologize in advance for these unintentional errors, and I hope they cause no harm.

Most of this book was written in the salubrious climate of the Côte d'Azur in the medieval perched village of Saint Jeannet, overlooking the Mediterranean Sea. I am forever indebted to Georges Carle, who made it possible for my family to stay in France over an eight-month period, to Patrick Gaudray, who invited me into his CNRS department at the Université de Nice, and to Claude Turc-Carel for graciously allowing me to invade her office one day each week. I also thank the patron of the Auberge du Quatre Routes for a daily *café double et un croissant*, and the other two thousand inhabitants of Saint Jeannet for their overflowing warmth and hospitality.

Finally, I feel it is important to acknowledge those I have never met who unknowingly made it easier to write this book. I thank the creators of the Macintosh computer system for making my research and writing a thoroughly enjoyable experience, and the creators of WordPerfect, Endnote Plus, Netscape, and Timbuktu software for providing tools for organizing my thoughts and allowing me to visit the whole world in a virtual sense from my tiny French village. And for soothing my spirit and forcing me to contemplate the meaning of consciousness, I thank Paul Simon, John Lennon, and Wolfgang Amadeus Mozart.

Notes

PROLOGUE: A GLIMPSE OF THINGS TO COME

p.   9   American society, in particular, accepts the rights of parents: In many ways, America is unique among Western countries in the paramount contribution that parents are expected to make in determining how their children should be raised. In most other industrialized countries, society as a whole takes primary responsibility for the socialization and upbringing of its children. Citizens of these countries would have greater legitimacy in arguing for control over the individual use of reprogenetic technologies.

1   WHAT IS LIFE?

p. 16   . . . all living entities on earth are . . . easily distinguished from nonliving things: Until 1828, it was thought that living matter was different in essence from nonliving matter. Living matter was considered to be "organic," nonliving matter was inorganic. Only living things, under divine influence, could produce organic materials. Thus, living things could be recognized because they were composed of, and produced, organic matter. This worldview was shattered in 1828 when the chemist Friedrich Wöhler synthesized the organic substance *urea* directly from inorganic materials in his laboratory. Scientists now use the word *organic* to describe the category of complex molecules that are based on the carbon atom. Although most of the complex molecules that living things produce are organic, these and many other organic molecules—never found in the living world—can also be synthesized in the laboratory with the techniques of organic chemistry.

p. 16 A different form of life, created: A lively account of the history, the players, and the field itself is provided by Steven Levy in *Artificial Life: the Quest for a New Creation* (New York: Pantheon, 1992).

p. 17 HAL was, of course, the vengeful computer: The name HAL was intended as a play on IBM, with each character advanced by a single position in the alphabet.

p. 17 HAL displayed unanticipated human emotions: The term artificial intelligence was coined in the 1950s to describe machines that might one day exhibit humanlike attributes of reasoning, self-awareness, and emotion.

p. 19 an absolute requirement for life of any kind is the ability to use energy: Some may quibble that the existence of spores contradicts this claim. Spores are dried-out cells that can survive in a completely inert state for hundreds of years and then germinate into living organisms when exposed to favorable environmental conditions. However, although spores emerge from living entities, and although they may have the potential to develop into living entities, they are better viewed as existing in a state of suspended animation. The rationale for this claim is presented in the discussion of frozen life in chapter 7.

p. 21 . . . evolution alone may be considered the overarching theme: It can be argued that the gene is a second essential feature of life-in-general. This argument follows directly from the principle that all living things are the products of reproduction and evolution. In theory, there are two different strategies a creature might take to reproduce itself. The first strategy would be for the creature to survey itself (not necessarily in a conscious way) from head to toe and use the structures of each of its component parts to produce identical copies in its emerging child. This is indeed the strategy that most biologists believe was used by the earliest living things on earth—self-replicating molecules that eventually evolved into the first biolife cell. It is the strategy that was used by the self-replicating robots in the story just told.

For the simplest living things, however, this strategy poses a serious problem in that only *self-replicating* component parts can be reproduced in this way. The requirement for self-replication severely restricts the ways in which a simple life form can evolve. For example, a membrane coat that formed, by chance, around a self-replicating molecule might increase the chances of that molecule's survival, but there is nothing that the self-replicating molecule could do to provide a similar membrane coat to its progeny, and thus the membrane coat would have no evolutionary significance. So with this strategy for reproduction, simple life forms wouldn't get very far. As a consequence, there could never be any self-replicating robots whose origin is dependent on intelligent creators. (Although they still provide the best counterexample to the argument that life requires genes.)

To get beyond the simplest stage of life, a creature has to be able to build structures to protect itself. But the best of these structures (and the machinery needed to build them) will not be self-replicating. So to make sure these structures are inherited, parents will need to give their offspring a set of *instructions* that tell them how they can also build these structures. This is the second strategy of reproduction. Give your children the instructions, with a minimal number of starting parts, and let them build themselves.

The second life strategy is very powerful in terms of both reproduction and evolution. If a creature needs one thousand identical parts of a particular surface structure (like an interlocking tile), they can all be constructed by reading a single instruction over and over. A change in this single instruction would lead to a change in all one thousand parts. If the changed overall structure of identical parts made the creature more resistant to attack or degradation than the original structure, the changed instruction would be passed on to more offspring, and eventually, all surviving members of the population would carry it. That's evolution.

The instructions carried by these hypothetical creatures are equivalent to what scientists call "genes." They are packets of *information* that are transferred from one generation to the next during the process of reproduction. Genes allow a child, in the most universal sense, to be created in the image of its parent or parents. The molecular substance that holds the genetic instructions in a form that can be both read and copied is called the *genetic material*.

Although the concept of the gene may be universal, the nature of the genetic material will not be. Every independently evolving system of life will have its own genetic material. Although DNA is a very beautiful molecule—both visually and conceptually—it is an accident. If we rewound the tape (as Stephen Jay Gould is fond of saying) back to Earth's beginnings and allowed life to evolve again *from scratch,* some other beautiful molecule would have emerged as the genetic material. So if a future solar system–hopping astronaut finds living things on another planet that happen to use DNA as their genetic material, she could be quite confident that the Panspermia Theory elucidated in chapter 2 is correct, and that we and these otherworldly creatures evolved from the same ancestral DNA-bearing life form whose descendants somehow migrated from one world to another.

p. 21 "Nothing in biology makes sense except in the light of evolution.": This aphorism is actually the title of a famous lecture given by Dobzhansky that was published in *The American Biology Teacher* 35 (1973): 125. It is often quoted and used as a rallying cry in defense of teaching evolution in the public schools.

p. 21 There seems to be a minimum level of complexity: While a-life creatures don't themselves obey this principle, they are dependent upon the complexity of the computer within which they reside to carry out their functions.

p. 21 Let's imagine that not only has artificial intelligence become a reality: The idea that scientists might be able to synthesize humanlike creatures, and the implications and consequences of such an idea, was first explored by Mary Shelley in her novel *Frankenstein*, written in 1816. Since Shelley's time, numerous writers have expounded upon this theme and its relevance to our understanding of the meaning of human life, as well as the separate but related notion of an ethical line beyond which science should not tread. An updated and poignant reconsideration of the Frankenstein theme was presented in 1921 by the Czech author Karel Čapek, who coined the word *robot* in a play called *R. U. R. (Rossum's Universal Robots): A Fantastic Melodrama.* Čapek's robots looked and acted like humans (rather than monsters) and thus paved the way for such modern film versions of the Frankenstein tale as *The Stepford Wives, The Terminator,* and *Blade Runner.*

p. 21 like an intelligent version of one of the Stepford Wives or a compassionate version of the Terminator: *The Stepford Wives* is the 1974 film adaptation of a novel by Ira Levin in which well-off businessmen in a sleepy Connecticut suburb of New York City kill their "demanding" spouses and replace each with a robotic copy that looks exactly like the original woman but that behaves according to a perfect model of a 1950s American housewife. This film has become a paradigm for the male desire to dominate the women in their lives and suppress any expression of their individuality.

Released in 1984, *The Terminator* is a film in which a robot with outwardly human characteristics (played by Arnold Schwarzenegger) is sent from the future into the present time in an attempt to change the course of history. *The Stepford Wives* and *The Terminator* both explore the premise of beings among us that look and act (to a certain extent) like humans when they are really not made of flesh and blood. Both films play upon the emotions of the audience in not immediately revealing the true nature of these creatures. Ultimately, the human look-alikes are revealed to lack certain human emotions, and in this way, both films pay homage to Shelley's *Frankenstein* with the implication that only God can provide the essential vital force that is the essence of humanness.

Many other films have explored the Frankenstein theme over the years, but *Blade Runner* was the first to challenge the view that only God can create human life. Although the naturally bred human characters in *Blade Runner* initially view the human-created "replicants" as less than human, the movie and the book on which it is based (*Do Androids Dream of Electric Sheep* by Philip K. Dick) force the audience to question this assumption.

2   WHERE DOES LIFE COME FROM?

p. 24   . . . one or more of these microscopic entities: You can get a sense of how large a typical cell is by comparing it to the distance between two of the closest markings on a standard American ruler. The typical cell would extend across only one hundredth the distance between two adjacent 1/16-inch lines.

p. 24   Thus, biolife cannot be reduced to any unit smaller than a cell: One question that may be asked is whether tiny viruses are also alive. Let's consider this question based on the properties of life-in-general that we've just discussed. A virus is essentially a strand of genetic material containing a handful of genes surrounded by a protective coat. It's hundreds to thousands of times smaller than a cell and cannot be seen with a standard microscope. In its natural state, the virus is completely inert. It can't use energy, it can't reproduce, it is absolutely static. It is only when the virus enters a cell and dissolves its coat that its exposed genetic material can take over the cell's machinery and force the production of more viruses—identical to itself—that are then exported to the world outside the cell. Thus, a virus sits on the very edge between living and inanimate. It can reproduce and evolve, but only within the helping environment of a cell. All viruses were born originally as escaped genes from cells, and all viruses are limited in the complexity they can achieve.

Of philosophical interest is the ability of scientists to create fully functional viruses in the laboratory, building them up from their smallest molecular components. When this feat was first accomplished in the 1970s, it provided the final proof that a so-called vital force was not necessary to explain life. Suddenly, it was no longer true that only life could beget life. Viruses will not be discussed further in this book because they are fringe entities that play no role in the definition of human life, reproduction, or development.

p. 24   It was assumed instead that a special *vital force:* The concept of the "vital force" emerged from the same tradition that had previously supposed an essential and "divine" difference between organic and inorganic matter. Once it became clear that molecules alone were not sufficient to distinguish the living from nonliving, the only thing left that could make living things different in essence was a special force that acted upon molecules to bring them to life. Hence the "vital force." Defined in different ways by different supporters, its underlying premise is that there is something present in every living cell and tissue—a guiding *spirit*—that transcends mere molecules. Although there is no need for a guiding spirit to explain the living cell, some still argue—on religious grounds—for its existence as an immaterial entity that does not interfere with the chemical processes of life. In this guise—containing no matter or energy itself, and unable to interact with matter—it ceases to exist within the confines of the rational world.

p. 25   . . . an exquisite piece of machinery with hundreds of thousands of working parts: Although the genetic material contains the instructions that a cell uses to build itself and its progeny, it doesn't actually carry out the work. Almost all the work that goes on inside the cell is carried out by molecules called proteins. Every living creature on Earth is composed primarily of protein molecules that can display almost infinite flexibility in their form and function. This is why early molecular biologists were so excited in the 1960s when they discovered the exact rules of the language that the cell uses to convert genetic information into functional proteins. Before I describe those rules (in a later note), it is important to understand a little bit about the chemical nature of proteins.

Chemists measure the size of molecules in terms of their *molecular weight,* which is based on a comparison to the size of the smallest possible atom with just a single proton and a single electron—hydrogen. Molecules of water, for example, contain two hydrogen atoms and a single oxygen atom, which together weigh 18 times as much as a hydrogen atom alone; thus a water molecule (H_2O) has a molecular weight of 18.

Living cells can make much bigger, and more complex, molecules by linking together large numbers of simple building blocks. The building blocks of proteins are called amino acids, which come in twenty different forms that range in molecular weight from 57 to 186. In 1953, the American biochemist Stanley Miller brought about the spontaneous creation of eight of these amino acids in a laboratory jar containing an atmosphere (with hydrogen, water, methane, and ammonia) and environmental conditions (electric discharges of the type caused by lightning) assumed to have existed on the newly formed earth, four billion years ago. This demonstration provided the first support for the idea that life could have emerged spontaneously on Earth.

The twenty amino acids have two parts—a shared backbone and a unique "residue." Amino acids are linked together through their shared backbones. It is their unique residue that defines their individual properties. Some amino acids possess electrical charge, either positive or negative; others are neutral. Some are slim, others are bulky. Some are flexible, others are taut. Some attract water molecules, others repel them.

A library of building blocks having such wonderfully diverse chemical properties allows the cell to create materials as different as the clear crystal lens in your eyes—which allows you to focus light on your retina—the strands of hair on your head, and the contractile substance inside your muscles; all these things are made primarily from proteins that differ only in the amounts and order of each amino acid they contain.

Some proteins have enzymatic activity, which means they can catalyze chemical reactions of many different kinds. It's this ability that allows proteins to function as the cell's machinery. In fact, it is proteins that act as the readers of genetic information. It is proteins that build new DNA molecules based on the information present in the old ones. It is proteins that build each component part of every cell in your body.

The cell builds proteins as long chains, connecting one amino acid to another through its backbone. A short chain of amino acids is often referred to simply as a peptide. A long chain (usually more than 100 amino acids, but the cutoff is arbitrary) is called a polypeptide. The smallest proteins produced by cells have just a handful of amino acids and act typically as molecular signals that can move quickly from one cell to another in the body. An average-size polypeptide contains about 500 amino acids and has a molecular weight of around 55,000. The largest known polypeptide is 3,685 amino acids in length with a molecular weight of more than 400,000; it is made in muscle cells and called dystrophin because mutations within it can lead to the disease of muscular dystrophy.

Proteins can actually contain multiple polypeptide chains as well as other small molecular structures. A good example is the hemoglobin protein made in red blood cells. Hemoglobin moves oxygen from your lungs to all the cells in your body. It has four polypeptide chains—two identical polypeptides called alpha-globin and two others called beta-globin—wrapped around a small iron-containing molecular module called a "heme." Genetically caused changes in either alpha- or beta-globin are responsible for the diseases of sickle cell anemia and thalassemia.

Although according to its scientifically agreed upon definition, a protein is larger than the individual polypeptide, scientists often use "protein" as a synonym for "polypeptide" (probably because "protein" has fewer syllables).

One last thing that you should know about proteins is that although they are created in one dimension—as chains of amino acids linked along a backbone—they become twisted and folded into highly specific three-dimensional structures. In fact, the vast majority of amino acids in most proteins are there solely to fit together with other specific amino acids—like a jigsaw puzzle—to ensure that proper twisting and folding occurs so that the proper shape and form of the protein is achieved. A protein must have a correct shape and form to function properly. Even a single amino acid change can destroy a protein's ability to reach its correct shape, with potential whole-body consequences like sickle cell anemia.

p. 25 The information . . . is encoded within its genetic material, DNA: The DNA molecules present in living cells are enormous compared to other kinds of molecules found in both the animate and inanimate worlds. An average-size human polypeptide—with a molecular weight of 50,000—may be enormous relative to molecules that form naturally in the absence of life. But even this pales in comparison to the size of an average molecule of human DNA with an approximate molecular weight of 80 billion.

If you could remove a DNA molecule from a cell and stretch it out without breaking it (actually a technical impossibility), you would find what looked like an incredibly long, but incredibly narrow, thread. All DNA molecules have the same diameter of about 2 nanometers, which is a million times smaller than the distance between two 1/16-inch marks on a ruler. This width is so small that naked DNA molecules cannot be detected with a traditional microscope.

The length of a typical human DNA molecule is another story. DNA molecules vary in length, but the average human molecule measures 1¾ inches. This is thousands of times longer than the human cell from which it was taken. And if the 46 DNA molecules present in each human cell were lined up end to end, they would measure an incredible six feet eight inches, longer than most human beings. But because their diameter is so narrow, an entire set of DNA molecules is easily compacted within each nucleus of each of the 100 trillion cells in your body.

There is a very good reason for the enormous difference between the width and length of a DNA molecule: genetic information is written across its length, not its width. A perfect analogy is obtained by considering what would happen if the text of the this entire book were written across a single line. The width of the line is only an eighth of an inch, but its length would be nearly a mile. The ratio of length to width in this case would be 470,000:1. For the average human DNA molecule, the ratio is 22,000,000:1. Although I've used an average human DNA molecule in the examples presented here, actual DNA molecules from humans and other species can vary greatly in length, just like books, in proportion to the amount of information they hold.

If you look at the total information contained on a page in this book, you will see that it is possible to divide it into smaller, well-defined units. There are paragraphs of various lengths, containing sentences of various lengths, each of which contains varying numbers of words of varying length. Finally, each word is composed of letters, or characters. The individual letter, or character, is the smallest bit of written information that cannot be broken down further. This is considered the *basic unit* of written information. When we give the size of word, we are not talking about its length in millimeters (which is arbitrary), we are talking about the number of letters, or basic units, it contains.

In a similar fashion, the information contained on magnetic disks used by computers can also be broken down into smaller units. The largest unit that a user sees is the file, which can vary in size over a very large range. Within a file, there are smaller units of identical size used for memory allocation or network transmission, which may be 256 bytes long. The byte has a standard size of 8 bits, and finally, the bit itself is the basic unit of computer information.

With both the written language and computer files, there is only a limited number of values that each basic unit can take. In the written information on this page, each basic unit takes the value of one of the twenty-six Roman letters or one of a limited number of auxiliary characters (such as a punctuation mark or blank space). In a computer file, each bit of information can take only one of two values—zero or one. But as we all know, this minimum number of alternative values does not limit, in any way, the amount of information that a whole file might contain.

The genetic language encoded within the DNA molecule also has a well-defined basic unit. This basic unit occurs within a modular component of DNA known chemically as a nucleotide. Since it represents the basic unit of genetic information, the nucleotide is also known simply as a *base*. There are only four different chemical bases, which can be used over and over again, and strung together in any order along a DNA molecule. The bases are referred to by the first letter of their chemical names: A (for adenine), C (for cytosine), G (for guanine), and T (for thymidine).

p. 26 All cells have two separate compartments: This statement is true only with qualifications. Bacteria are primitive forms of cells that are not compartmentalized; their genetic material floats freely in their cytoplasm. And there are some specialized cells in complex creatures like us—such as red blood cells—that throw away their nuclei and end up functioning perfectly well with only a cytoplasm (although such cells are unable to grow or divide).

p. 26 . . . the cells of all living things work in essentially the same way: The information required to produce the three-dimensional structure of a protein lies entirely within the one-dimensional sequence of amino acids present in its polypeptide chain. This one-dimensional sequence of amino acids is determined entirely by a packet of information present along a DNA molecule: the gene. There is a one-to-one correspondence between genes and polypeptides. But a polypeptide looks nothing like a DNA molecule. A DNA molecule has only four building blocks—A,C,G, and T—while a polypeptide has twenty. This means that the transfer of information from a sequence of DNA bases—a *DNA sequence*—to a sequence of amino acids—a *polypeptide* or *amino acid sequence*—is not accomplished through the operation of a simple one-to-one translation mechanism. Instead, the polypeptide sequence must be encoded somehow within the gene sequence. But how? What is the *genetic code*?

After the 1953 determination of the structure of DNA by James D. Watson and Francis H. Crick, molecular biologists worked for another dozen years to unravel the genetic code. When the discovery was finally made (it actually occurred through a series of smaller discoveries), it was seen as a great breakthrough in human understanding. Here it is in a summary version.

Each amino acid in a polypeptide is coded for by a specific sequence of three DNA bases. This three-base sequence unit is called a *codon*. Since each DNA base can take one of four forms, a codon can take one of $4 \times 4 \times 4$ or 64 forms. Thus, there is a potential for 64 different codons, which is about 3 times the number of different amino acids. On average, 3 different codons will specify a single amino acid. For example, the codons ATT, ATC, and ATA all specify an amino acid named isoleucine. But this number can vary from as low as 1 codon (for each of 2 amino acids) to as high as 6 codons (for each of 2 other amino acids).

Of the 64 possible codons, 61 specify one amino acid or another. The other three codons (TAA, TAG, and TGA) represent a punctuation mark signifying the end of a gene; these are called STOP codons. The DNA *coding region* for every polypeptide starts with ATG, which is referred to as a START codon. Thus each gene extends from START to STOP. ATG encodes the amino acid methionine, which is at the beginning of every polypeptide. The precise order and number of amino acids in each polypeptide chain are determined by the codons that begin with START and continue to STOP. Each codon follows directly after the preceding one, so the size of the coding region, in bases, is always three times the number of amino acids in the polypeptide. That, in essence, is the whole story of the genetic code.

p. 27 The probability is 1 chance in 403,291,146,110,000,000,000,000,000: The number of different ways in which the individual symbols of a twenty-six-character alphabet could be assigned to twenty-six different sounds is calculated with a string of multiplications starting with 26 and moving down by one with each subsequent number: 26 × 25 × 24 . . . 2 × 1. This mathematical expression is referred to as 26 factorial and is written as *26!*

p. 27 And yet, every living animal, plant, and germ cell on earth: There are some minor differences in the code used by certain microorganisms. But even these exceptional organisms still retain more than 90 percent of the code used by our own cells, which tells us that these differences occurred *after* the birth of the first cell.

p. 27 . . . powerful enough together to change . . . the chemical nature of its atmosphere: The most striking of these transformations was the conversion of an atmosphere rich in carbon dioxide and nearly lacking in oxygen to one at the opposite extreme. This drastic conversion was a consequence of photosynthesis being carried out by trillions and trillions of primitive green cells that coated the surface of our planet billions of years ago. During photosynthesis, cells use carbon dioxide and water in the generation of chemical energy, while excreting oxygen as a waste product. It is the accumulation of this waste product that allowed the evolution of animal life, which eventually led to us.

p. 29 . . . the first replicators were protein-like in nature: S. L. Miller, "The Prebiotic Synthesis of Organic Compunds as a Step Toward the Origin of Life," in *Major Events in the History of Life*, ed. J. W. Schopf (Boston: Jones & Bartlett, 1992), pp. 1–28.

p. 29 . . . RNA was the original replicator: W. Gilbert, "The RNA World," *Nature* 319 (1986): 618.

p. 31 Francis Crick . . . has tried to rescue Panspermia: This idea is the premise for Crick's book *Life Itself: Its Origin and Nature* (New York: Simon and Schuster, 1981).

p. 31 . . . Pope John Paul's decision to accept evolution as an established scientific theory: As reported in the *New York Times* on 25 October 1996, the Pope said, "Fresh knowledge leads to recognition of the theory of evolution as more than just a hypothesis."

p. 32 . . . a problem that has been addressed most forcefully: Stephen Jay Gould, *Wonderful Life: The Burgess Shale and the Nature of History* (New York: W.W. Norton, 1989).

p. 32 "The more the universe seems comprehensible,": Steven Weinberg, *The First Three Minutes* (New York: Bantam Books, 1977).

p. 32 "Matter is weird stuff": Freeman Dyson, *Infinite in all Directions* (New York: Harper & Row, 1988), p. 8.

p. 33 . . . a backhand way to explain the particular properties: John D. Barrow and Frank J. Tipler, *The Anthropic Cosmological Principle* (New York: Oxford University Press, 1986); Martin Rees, *Before the Beginning: Our Universe and Others* (New York: Simon & Schuster, 1997).

p. 33 Ed Witten . . . rejects this version of the Anthropic Principle: Witten says, "What I detest most about the Anthropic Principle is that it robs us of the real wonder, which is that after learning conventional (nonanthropic) explanations of things, we find out that for 'purely theoretical' or 'purely mathematical' reasons, things turn out to be just so as to make life possible. It is very roughly as if we would determine that life would be possible only if pi is between 3.1415 and 3.1416 and then, upon making accurate measurements of the circumference of a unit circle, learn that that is true. I see the Anthropic Principle as an excuse to surrender the quest for a scientific explanation. Such surrender always turns out, in the long run, to be misguided." (Personal communication, 1997).

p. 33 . . . a new theory of physics that brings together the currently separate theories: The reigning theories of physics are the Quantum Mechanical "Standard Model" and Einstein's General Theory of Relativity. The Standard Model does a good job of explaining the electrical and nuclear forces, while Relativity does a good job of explaining gravity. But the two theories collide in their descriptions of what happens when two elementary particles come very close to each other, which is why physicists know something is missing. Furthermore, neither theory explains why elementary particles have the masses that they have, or why forces have the precise strengths that they have. These fundamental constants are simply observed in nature and incorporated into the theories. Physicists such as Ed Witten and David Gross hope that a grand new mathematical construct called "String Theory" will bring together quantum mechanics and gravity and provide an explanation for the fundamental constants.

p. 34 . . . "the laws of Nature and the initial conditions": Dyson, *Infinite,* p. 298.

p. 34 . . . "small piece of God's mental apparatus": Ibid., p. 297.

p. 34 I refuse to believe that knowledge exists: For a point of view that is the polar opposite of mine, the reader is directed to a book by John Horgan called *The End of Science* (Reading, Mass.: Addison-Wesley, 1996).

3 DOES YOUR FIRST CELL DESERVE RESPECT?

p. 35 What is the moral status of the human embryo?: An excellent analysis of this question from a political perspective is presented by Alta Charo in an article titled, "The Hunting of the Snark: The Moral Status of Embryos, Right-to-Lifers, and Third World Women," *Stanford Law and Policy Review* 6 (1995): 11–37.

p. 36 The two members of any chromosome pair are 99.9 percent equivalent to each other: The one exception lies in the pair of sex chromosomes. Unlike the other twenty-two pairs, sex chromosomes come in two different forms: X and Y. Women have a pair of X chromosomes which are 99.9 percent the same as each other, just like other chromosome pairs. Men have a single X and a single Y with very different genetic information.

p. 38 . . . the two pronuclei never fuse into one: Personal communication from Alan Trounson, Centre for Early Human Development, Monash Medical Center, Clayton, Australia, on 20 June 1996.

p. 39 So why is there suddenly a need: As of this writing, the contemporary meaning of the word *pre-embryo* has not been included in the on-line versions of the *Oxford English Dictionary* or the *Merriam-Webster's Collegiate, Tenth Edition,* but is included in the 1992 version of the *American Heritage Electronic Dictionary.*

p. 39 The term pre-embryo has been embraced: The eminent biochemist Erwin Chargaff, who has long railed against the use of genetic engineering in all biological realms, had this to say shortly after the introduction of the term: "The 'pre-embryo' is a designation that appears to me entirely unjustified. I fear that it is merely an alibi function . . . The attempt to determine, by scientific means, the stage at which what for times immemorial had been called the human soul makes it appearance, is ridiculous. The setting of a calendar date serves only as a permit for the performance of experiments that normal reverence before human life would have outlawed . . ." Erwin Chargaff, "Engineering a Molecular Nightmare," *Nature* 327 (1987): 199–200.

p. 40 . . . the first committee ever commissioned by a government: In 1982, the British Parliament established a committee for the purpose of examining "the social, ethical, and legal implications of recent and potential developments in the field of assisted human reproduction." The Warnock Committee—as it came to be known because of its chairperson, Mary Warnock—issued its report on 25 June 1984. The report, with its sixty-three separate procedural and legal recommendations, had an enormous impact on the way reproductive technologies were viewed not only in Great Britain but around the world. The Warnock Committee was the first high-level governmental commission to mention—let alone make recommendations on—in vitro fertilization, egg and embryo donation, embryo freezing, trans-species fertilization, nonhuman surrogate mothers, and many other more mundane reproductive technologies.

p. 41 "The embryo deserves *respect* greater than that accorded": John Robertson, *Children of Choice: Freedom and the New Reproductive Technologies* (Princeton, N.J.: Princeton University Press, 1994), p. 102.

p. 42 "modern genetic science brings valuable confirmation.": Vatican Congregation for the Doctrine of the Faith, "Instruction on Respect for Human Life in Its Origin and on the Dignity of Procreation," *Origins, NC Documentary Service* 16 (1987): 697–711. (*Origins, NC Documentary Service* is published weekly by the National Catholic News Service, Washington, D.C.) The complete text of the original document was also published in the *New York Times* (11 March 1987), and in an abridged form in *The Ethics of Reproductive Technology*, ed. Kenneth D. Alpern (New York: Oxford University Press, 1992), pp. 83–97. The parenthetical phrase in the *Origins* publication was transformed into a footnote (on page A14) of the *New York Times* version, which reads as follows: "The zygote is the cell produced when the nuclei of the two gametes have fused."

p. 42 "While the egg and sperm are alive as cells": Leon Kass, "The Meaning of Life—In the Laboratory," in *Ethics*, ed. Alpern, pp. 98–116.

p. 43 But even at this stage: Paternal and maternal DNA molecules only interact and touch each other within the germ cells that lead to the formation of the *next* generation of sperm or eggs. In males, this doesn't occur until puberty. In females, it occurs in the ovaries of the midgestation fetus.

p. 44 If a human life can begin in the absence of conception: Inversely, if conception does mark the beginning of all new human lives, then human beings cannot be cloned. While there is no scientific support for this alternative point of view, for the time being, it is the only conclusion that can be reached by those who refuse to give up their faith in the "life begins at conception" principle. This idea was proposed in an essay that I received by mail from Fr. Anthony Zimmerman of Nagoya, Japan. Mr. Zimmerman believes that there is no need to talk about the implications of human cloning because it can never happen.

p. 44 And it's the *feelings* of animals that we respect: The issue of respect for living things is somewhat more complicated than I have presented here. One can imagine having respect for the survival of a particular *species* of animal or plant, or for a more complicated combination of living things that make up a particular ecosystem. This type of respect transcends the individual organism and resembles the kind of respect that we give to inanimate objects perceived as symbols, as will be discussed shortly.

p. 45 There are two given by Robertson: Robertson, *Children of Choice*, p. 102.

p. 45   . . . millions of sperm in an ejaculate: The absurd notion that even sperm cells deserve respect is parodied by Monty Python in the song "Every Sperm Is Sacred," written by Michael Palin and Terry Jones and performed in the movie *The Meaning of Life*.

p. 46   the Vatican has said that prenatal diagnosis is morally permissible: *Origins*, p. 702.

p. 46   "embryos are potent symbols of human life": Robertson, *Children of Choice*, p. 252.

p. 47   "the preimplantation embryo warrants serious moral consideration": Ad Hoc Group of Consultants to the Advisory Committee to the Director of the NIH, *Report of the Human Embryo Research Panel, Vol. 1* (NIH publication number 95–3916, September 1994), p. x.

4   FROM YOUR FIRST CELL TO YOU

p. 49   . . . to make an informed decision on the question of the emergence of human life: My own response to the question, When does human life emerge? has been greatly informed by the scientific arguments presented in Harold Morowitz and James Trefil, *The Facts of Life* (New York: Oxford University Press, 1992).

p. 49   The embryo still looks like a ball, or rather, a microscopic raspberry: This image was first suggested by Robert Edwards, one of the co-inventors of IVF, in his book *Life Before Birth: Reflections on the Embryo Debate* (London: Hutchinson, 1989), p. 50.

p. 49   Once a cell has undergone differentiation: A critical and fascinating exception to this rule occurs with the onset of some cancers when differentiated cells can revert to an earlier stage. This process of "de-differentiation" is caused by mutations in the cell's genetic material that can occur long after a person is born. As we now know (since February 1997), a second, incredible exception is possible through the fusion of a differentiated cell with an unfertilized, nuclear-free egg. This process appears to allow the complete reversal of the differentiated state.

p. 50   . . . a differentiated cell doesn't lose any of its genetic information: This statement is almost universally true, but not quite. In certain cells of the immune system, tiny pieces of DNA are spliced out of particular chromosomal regions for the purpose of bringing together the building blocks of new genes that are unique to each individual cell. In addition, red blood cells eject their entire nuclei during the process of terminal differentiation.

p. 51 For women who use an Intrauterine Device (IUD): The IUD is inserted by a physician directly into a woman's uterus, where it can remain for a long time. The IUD does not block fertilization. Instead, it interferes with the implantation of the embryo into the uterine wall.

p. 53 "Probably, most people unfamiliar with this field": C. R. Austin, *Human Embryos: The Debate on Assisted Reproduction* (Oxford: Oxford University Press, 1989), p. 17.

p. 55 It is during this period that the fetus develops: There is a small number of reports of survival outside the womb at or before twenty-two weeks of gestation. Based on our understanding of fetal development, it is almost certain that the age of these fetuses was simply misdiagnosed, and they were older than they seemed at the time of their removal from the uterus.

p. 56 It is certainly possible, if not likely: Some progress toward the ultimate goal of ectogenesis—the word used to describe gestation outside the human body—has been made by a team of Japanese scientists who have succeeded in keeping goat fetuses alive for three weeks in an incubator filled with artificial amniotic fluid. But this work represents only a tiny step toward the creation of a full-term artificial womb. The scientific reference for this work is N. Unno, Y. Kuwabara, T. Okai, K. Kido, H. Nakayama, et al., "Development of an Artificial Placenta: Survival of Isolated Goat Fetuses for Three Weeks with Umbilical Arteriovenous Extracorporeal Membrane Oxygenation," *Artificial Organs* 17 (1993): 996–1003. A news report of this work appeared in a story by Perri Klass entitled "The Artificial Womb Is Born" in the *New York Times*, 29 September 1996.

p. 56 *Most brain cells are produced early in the pregnancy*: Morowitz and Trefil, *Facts*, p. 117.

p. 57 "humanness and the ability to survive . . .": Ibid., p. 146.

p. 57 The average time from conception to birth: Physicians often time a pregnancy from the first day of a woman's last menstrual period. Since conception usually occurs about two weeks after the start of the menstrual cycle, the average length of a pregnancy measured in this way is 40.5 weeks, rather than 38.5. Alan F. Guttmacher and Irwin H. Kaiser, *Pregnancy, Birth and Family Planning* (New York: E. P. Dutton, 1984).

p. 58 . . . it is possible to conclude that you didn't exist: Based on the notion of consciousness as the defining feature of human life, the philosopher Michael Tooley has argued that to maintain logical consistency, the newborn infant should be granted no more right to life than an embryo or fetus. Based on this point of view, Tooley argues that early infanticide is morally acceptable. Michael Tooley, *Abortion and Infanticide* (New York: Oxford University Press, 1983.)

5 BABIES WITHOUT SEX

p. 65 . . . the Vatican seems willing: This variation is described as follows in a sidebar to the previously cited *Origins* article: "Some Catholics have judged the technique [GIFT] acceptable provided that masturbation is not involved in collection of the sperm. A perforated condom is used during intercourse, with the sperm retrieved from the condom afterward . . . Monsignor Elio Sgreccia [a Roman Catholic priest and ethicist who contributed to the cited article] . . . said methods that seek to help marital intercourse attain fertility should be considered 'within the range of licitness.' . . . It is hoped that science would make available other fertility techniques that retain the conjugal act as the source of life and help it reach its full effect, Sgreccia said." Vatican Congregation for the Doctrine of the Faith, "Instruction on Respect for Human Life in Its Origin and on the Dignity of Procreation," *Origins, NC Documentary Service* 16 (1987): 699.

In an article in the *New York Times* by James Gleick (12 March 1987), the rationale for the perforated condom is explained: "Catholic medical authorities said sperm for such procedures could be obtained ethically by using a condom deliberately pierced with holes to allow some sperm to escape. A standard condom is unacceptable to the Church, since it is a means of contraception, even if the ultimate goal is to conceive a child."

Of course, *contraception* has no meaning in the context of an infertile couple since *conception* will not occur whether the holes are there or not. This leads to the question of how big the holes must be. If one sperm gets through, is that enough? Or must the holes be large enough to allow tens, hundreds, or thousands of sperm to get through? No matter how many get through, conception can't possibly occur in a woman with blocked fallopian tubes. Then there's the added complication that different subfertile men are likely to have widely different concentrations of sperm in their semen.

p. 66 In contrast, when women conspire to bring about fetal development: In a similar vein, the Catholic philosopher Oliver O'Donovan has written, "I confess that I do not know how to think of an IVF child except . . . as the *creature* [original emphasis] of the doctors who assisted at her conception." Oliver O'Donovan, "Begotten or Made?" in *The Ethics of Reproductive Technology,* ed. Kenneth D. Alpern (New York: Oxford University Press, 1992), pp. 195–202.

p. 66 "reproductive technologies . . . are transforming": Gena Corea, *The Mother Machine* (New York: Harper & Row, 1985), pp. 288–90.

6 IN VITRO FERTILIZATION AND THE DAWN OF A NEW AGE

p. 68 Robert Edwards—the original figure: Edwards, *Life Before Birth*, p. 126.

p. 68 . . . techniques perfected by other mouse embryologists: The first documented transfer of embryos from one mammal to another was reported in 1890 by Walter Heape in a paper entitled "Preliminary Note on the Transplantation and Growth of Mammalian Ova within a Uterine Foster-Mother," *Proceedings of the Royal Society* 48 (1890): 457–58. Heape transferred embryos that had resulted from a mating between two "Angora" rabbits into the fallopian tube of a "Belgian hare." The Belgian hare gave birth to two rabbits with all of the unique characteristics of the Angora breed including albino fur color, "long silky hair peculiar to the breed," and "a habit of slowly swaying their head from side to side as they look at you." In effect, the Belgian hare that Heape experimented on represents the first mammal ever to have given birth to offspring that were not genetically her own.

p. 68 The problem was not in getting fertilization: Edwards and Steptoe first reported their success at human in vitro fertilization in an article entitled "Early stages of fertilization in vitro of human oocytes matured in vitro," published in *Nature* 221 (1969): 632. A quarter of a century earlier, John Rock and Miriam Menkin, working at Harvard Medical School, claimed to have accomplished this task ("In Vitro Fertilization and Cleavage of Human Ovarian Eggs," *Science* 100 (1944): 105–107), but modern methods of analysis suggest that they may have misinterpreted their data.

p. 69 In 1985, the first year that a survey of U.S. clinics was performed: Statistical data compiled from all registered IVF centers in the United States by the Society for Assisted Reproductive Technology (SART) are published each year by the American Society for Reproductive Medicine (ASRM) in the journal *Fertility and Sterility*. Summary information is also available on World Wide Web pages maintained by ASRM. Their web site address is http://www.asrm.com. To purchase written reports, contact the ASRM at 205–978–5000.

p. 69 By 1994, more than thirty-eight countries had established IVF programs: Information on the worldwide use of IVF is extracted from a survey conducted by the pharmaceutical company Organon in 1994 and available on the World Wide Web at http://www.bris.ac.uk/Depts/ObsGyn/crm/ivf94.html.

p. 69 At a typical IVF clinic: P. J. Neumann, et al., "The Cost of a Successful Delivery with In Vitro Fertilisation," *New England Journal of Medicine* 331 (1994): 239–43 and 270–71.

p. 70 The desire to have and raise a child: In the essay "Genetic Puzzles and Stork Stories: On the Meaning and Significance of Having Children," the philosopher Kenneth Alpern explores the broad range of answers given by people who are asked why they want to have children, and the so-called validity of each answer. *The Ethics of Reproductive Technology*, ed. Kenneth D. Alpern (New York, Oxford University Press, 1992), pp. 147–69.

p. 70 Actually, they all do, otherwise they wouldn't be present: Although I've made the statement that all our genes play a role in making us "fit" from a Darwinian perspective, this may not be entirely true according to some models of evolution. There are hundreds of thousands of genetic segments within our chromosomes that appear to function for their own benefit rather than for the benefit of the organism within which they lie (meaning you and me). These genetic segments are referred to as "selfish DNA" when they are small, or "selfish chromosomes" when they are large (not to be confused with the "Selfish Gene," a term coined by Richard Dawkins and the title of the book in which he presents the view that *all* genes—by definition—act in a manner that is fundamentally selfish).

Readers who would like to learn more about selfish DNA and selfish chromosomes can turn to review articles by the author and others: L. M. Silver, "The Peculiar Journey of a Selfish Chromosome: Mouse t Haplotypes and Meiotic Drive," *Trends in Genetics* 9 (1993): 250–54. L. E. Orgel and F. H. C. Crick, "Selfish DNA: The Ultimate Parasite," *Nature* 284 (1980): 604–607. W. F. Doolittle and C. Sapienza, "Selfish Genes, the Phenotype Paradigm and Genome Evolution," *Nature* 284 (1980): 601–603.

p. 71 One of these is an innate fear of snakes: Edward O. Wilson, *In Search of Nature* (Washington, D.C.: Island Press, 1996), pp. 18–30.

p. 71 Ultimately, the emotional desire to have children: There are still gene critics in the social sciences who refuse to accept the idea that the human desire to have children is instinctual. They claim instead, "the notion that a desire for children is natural and instinctive might also be considered a nonconscious ideology," which is based on a "social construct." H. B. Holmes, *Issues in Reproductive Technology* (New York: Garland Publishing, 1992), p. 271. In other words, the *only* reason people want to have children is because society makes them feel that way without their realizing it. Those who spout such nonsense clearly lack even a modicum of understanding of evolution in general, or the genetic contribution to behavioral predispositions in particular.

p. 71 For the vast majority of people: According to a 1990 Gallup poll, 84 percent of childless adults under the age of forty would like to have children, and 60 percent of childless adults aged forty or older wish they had children.

p. 73 And the overall success rate: It is not possible to come up with a single figure to describe the rate at which IVF succeeds as a treatment for infertility. Since the IVF industry in the United States is totally unregulated, any licensed physician can set up a private IVF clinic. As a result, there are many clinics with outdated equipment or less than fully competent practitioners who achieve very low rates of success (even as low as 0 percent). National surveys do not distinguish between good and bad clinics, and as a consequence, the average success rates reported are low and quite meaningless.

Even at the best clinics, the rate of success will vary tremendously according to the nature of the fertility problem, if any; the age of the potential parents; and many other factors. In the best circumstances, when IVF is used simply as means for selecting nondiseased embryos (as described in chapter 17) for introduction into a young fertile woman, the rate of successful pregnancy and birth can be as high or higher than that achieved naturally.

p. 73 Furthermore, the rapid-response events: M. Maleszewski, Y. Kimura, and R. Yanagimachi, "Sperm Membrane Incorporation into Oolemma Contributes to the Oolemma Block to Sperm Penetration: Evidence Based on Intracytoplasmic Sperm Injection Experiments in the Mouse," *Molecular Reproduction and Developmental Biology* 4 (1996): 256–59.

p. 74 The process is called Round Spermatid Nucleus Injection or ROSNI: The IVF field is loaded with acronyms. It seems that any improvement in IVF technology, no matter how insignificant, is given its very own name. Many fall quickly by the wayside, but others become established terms. In addition to IVF, ICSI, and ROSNI, other established acronyms include GIFT (Gamete Intrafallopian Transfer), ZIFT (Zygote Intrafallopian Transfer), TET (Tubal Embryo Transfer), PZD (Partial Zona Dissection), and SUZI (Sub-Zonal Insertion). Another acronym, ART—for Assisted Reproductive Technology—is the umbrella term that the profession itself uses to encompass all of the various technical variations upon the IVF concept. To most members of the general public, however, it's all IVF. And my own feeling, as well, is that IVF represents the concept of in vitro fertilization rather than any particular technique. So, IVF is the term that I use in this book to refer to any protocol that involves fertilization in the laboratory followed by embryo transfer back into a living body.

p. 74 Ralph Brinster, at the University of Pennsylvania: R. L. Brinster and J. W. Zimmermann, "Spermatogenesis Following Male Germ-Cell Transplantation," *Proceedings of the National Academy of Sciences* 91 (1994): 11298–302.

p. 74 . . . "change the nature of our species": editorial entitled "Exploring Life as We Don't Yet Know It," *Nature* (7 March 1996): 89.

p. 75 More than 75 percent of Americans now feel: Results of a survey conducted for *Family Circle* magazine by Princeton Survey Research Associates in May 1994.

p. 75 "Any change in custom": Sophia J. Kleegman and Sherwin A. Kaufman, *Infertility in Women* (Philadelphia: F. A. Davis, 1966), p. 178.

p. 76 . . . a hypothetical scenario in which genetic engineering: The use of genetic technology to provide immunity to infection by the AIDS-causing virus HIV is no longer just hypothetical. In 1996, researchers demonstrated the existence of a small proportion of men who were HIV-resistant because they carried a particular form (allele) of a particular gene. This information could be used to genetically engineer embryos so that the children who emerged would, indeed, be HIV-resistant as well.

7   FROZEN LIFE

p. 79 Zoe Leyland became the first human child: "The New Origins of Life," *Time,* 10 September 1984, p. 40.

p. 81 "a solution to the collection of excess oocytes": Alan Trounson, "Preservation of Human Eggs and Embryos," *Fertility and Sterility* 46 (1986): 1–12.

p. 87 . . . hundreds of women in Italy: Quoted in article by Fred Barbash in the *Washington Post,* "British Law to Thaw 3,000 Embryos," 1 August 1996.

8   FROM SCIENCE FICTION TO REALITY

p. 92 Ninety percent of Americans polled: Data extracted from a *Time*/CNN poll taken over 26 and 27 February 1997, and reported in *Time* on 10 March 1997; also an ABC *Nightline* poll taken over the same period, with results reported in the *Chicago Tribune* on 2 March 1997.

p. 92 . . . was called "morally despicable," "repugnant," "totally inappropriate,": Quotes from the bioethicist Arthur Caplan in *The Denver Post,* 24 February 1997; the bioethicist Thomas Murray in the *New York Times,* 6 March 1997; Congressman Vernon Elders in the *New York Times,* 6 March 1997;

and evolutionary biologist Francisco Ayala in *The Orange County Register*, 25 February 1997.

p. 92   Most unhappy of all were those associated: James A. Geraghty, president of Genzyme Transgenics Corporation (a Massachusetts biotech company) testified before a Senate committee that "everyone in the biotechnology industry shares the unequivocal conviction that there is no place for the cloning of human beings in our society." (*Washington Post*, 13 March 1997).

p. 92   . . . two out of three Americans: Data obtained from a Yankelovich poll of 1,005 adults reported in the *St. Louis Post-Dispatch* on 9 March 1997 and a *Time/CNN* poll reported in the *New York Times* on 5 March 1997.

p. 92   . . . it might not be possible *at all* to transfer the technology: Leonard Bell, president and chief executive of Alexion Pharmaceuticals, is quoted as saying, "There is a healthy skepticism whether you can accomplish this efficiently in another species." In the *New York Times*, 3 March 1997.

p. 93   . . . "it would take years of trial and error": Interpretation of the judgments of scientists reported by Michael Specter and Gina Kolata in the *New York Times*, 3 March 1997, and by Wray Herbert, Jeffrey L. Sheler, and Traci Watson in *U.S. News & World Report*, 10 March 1997.

p. 93   . . . "there is no clinical reason why you would do this": Quote from Ian Wilmut, the scientist who brought forth Dolly, in article by Tim Friend for *USA Today*, 24 February 1997.

p. 93   The direct injection of sperm into eggs (ICSI): The main concern with ICSI was based on the notion that, in the *natural* process of fertilization, genetically healthy sperm beat out genetically unhealthy sperm. Following this line of reasoning, if the reprogeneticist instead chooses the sperm for fertilization, competition among sperm is not allowed to occur and a higher percentage of "genetically defective" sperm will take part in the formation of embryos. Although a number of prominent physicians express this concern, there is no scientific data, or even a conceptual framework, to support it. On the contrary, the vast majority of new mutations will have no effect on a sperm's ability to carry out fertilization. Consequently, sperm that carry mutations will behave no differently than sperm that don't carry mutations. But it is still the case that the safety of any new medical technology can never be determined in the absence of experimental data.

p. 93   The word *clone*: H. J. Webber first used the word *clon* in the 16 October 1903 issue of *Science*, p. 502. C. L. Pollard modified the spelling to *clone* in the 21 July 1905 issue of *Science*, p. 88. The word *clone* is derived from *klon*, the Greek word for twig.

p. 94   Among single-cell organisms like bacteria: Indeed, to molecular biologists and microbiologists, the word *clone* brings to mind a colony of bacteria rather than a colony of people. And since 1975, the word has had a special meaning in the context of a "DNA clone" or "gene clone." In these terms, a clone refers to the millions or billions of identical copies of a specific fragment of human or mouse DNA, for example, generated in a clone of bacterial cells through the use of recombinant DNA technology.

p. 95   The use of "nuclear transplantation": R. Briggs and T. J. King, "Transplantation of Living Nuclei from Blastula Cells into Enucleated Frogs' Eggs," *Proceedings of the National Academy of Sciences USA* 38 (1952): 455–63.

p. 96   . . . John Gurdon, who finally succeeded: J. B. Gurdon, "Transplanted Nuclei and Cell Differentiation," *Scientific American* 219 (1968): 24–35.

p. 96   . . . "the cloning of mammals by simple nuclear transfer": J. McGrath and D. Solter, "Inability of Mouse Blastomere Nuclei Transferred to Enucleated Zygotes to Support Development In Vitro," *Science* 226 (1984): 1317–19.

p. 96   The idea began to filter into the public consciousness: The rapid change in the public perception of a clone can be seen in a science fiction novel published in 1965 that was entitled simply *The Clone*, by T. L. Thomas and K. Wilhelm (New York: Berkley, 1965). In this book, the clone is an organism that replicates itself asexually in the classical sense of the word as it was originally used for colonies of bacteria or yeast. There was no indication in this book that a clone could possibly be used to describe people or even animals. And yet, within less than a decade, the public perception of a clone became directly focused on human beings.

p. 97   "One of the more fantastic possibilities": Alvin Toffler, *Future Shock* (New York: Random House, 1970), p. 197, from the Bantam paperback edition.

p. 97   . . . several years later his publisher was forced: Although his publisher was forced to capitulate in response to a lawsuit, Rorvik himself never wavered from his claim that what he had written was true. Two decades later, the issue was revived with the announcement of Dolly's birth. Writing for the June 1997 on-line issue of *Omni* magazine (http://www.omnimag.com), Rorvik now sees his vindication in the demonstration that mammalian cloning really is possible and relatively easy to accomplish. Most scientists, however, still believe that the technical expertise and knowledge was simply not available for cloning to be carried out successfully in 1978.

p. 98   . . . a report that two George Washington University scientists: J. L. Hall, D. Engel, G. L. Motta, P. R. Gindoff and R. J. Stillman, "Experimental

Cloning of Human Polyploid Embryos Using an Artificial Zona Pellucida," *American Fertility Society program supplement: Abstracts of the Scientific Oral and Poster Sessions* S1 (1993).

p. 98 The experiment was terminated at this point: Actually, the scientists had purposely chosen to use embryos that had undergone fertilization with two sperm and, as such, would not be viable for more than a few days. They had assumed that by experimenting on such embryos—which have no potential to reach even the earliest stages of fetal differentiation—they could avoid ethical concerns. Their experiment was preapproved by their hospital's ethical review board, whose chairperson, Gail Povar, later said, "What we're talking about here is nonviable human chromosomal tissue. I don't consider this to be human cloning. I consider it the manipulation of pathological specimens." (Quoted in *New Scientist,* 30 October 1993, p. 7).

In addition, it is important to point out that many other scientists in the field of reproductive biology saw nothing original in the work performed by Hall and Stillman. Techniques for splitting animal embryos were well established many years earlier, and techniques for separating cells of the human embryo were also commonly practiced by this time as a means for obtaining biopsy material for genetic diagnosis as described in chapter 17. The only addition made by Hall and Stillman was to allow multiple separated cells from the same embryo to remain alive and undergo further divisions in a laboratory dish over a period of a few days.

p. 98 But even the normal practice of IVF: Of the 6,870 children born by IVF in the United States in 1993, 65.9 percent were singletons, 27.5 percent were twins, 5.4 percent were triplets, and 0.4 percent were quadruplets or quintuplets, as reported in the July 1995 issue of *Fertility and Sterility.*

p. 98 The Vatican called it a "perverse choice": Article entitled "Cloning: Where Do We Draw the Line?" by Philip Elmer-Dewitt in *Time,* 8 November 1993, p. 64.

p. 98 Biotech critic Jeremy Rifkin: Ibid.

pp.98– "Scientist Clones Human Embryos and Creates an Ethical Challenge":
99 The first news report, with this title, was written by Gina Kolata and published in the *New York Times* on 24 October 1993.

p. 99 The first step was accomplished: J. McGrath and D. Solter, "Nuclear Transplantation in the Mouse Embryo by Microsurgery and Cell Fusion," *Science* 220 (1983): 1300–02.

p. 99 The next advance on the way to Dolly: S. M. Willadsen, "Nuclear Transplantation in Sheep Embryos," *Nature* 320 (1986): 63–65.

p. 100   . . . an unfertilized egg is chock-full of signal proteins: The assumption is that the critical signals are no longer present in the *cytoplasm* of a fertilized egg because they have already jumped onto the sperm DNA that previously entered the cell.

p. 100   Eight more years went by: N. L. First and M. Sims (1994) "Production of Calves by Transfer of Nuclei from Cultured Inner Cell Mass Cells," *Proceedings of the National Academy of Sciences, USA* 90: 6143–47.

p. 100   . . . a technician in First's laboratory: Reported by Michael Specter and Gina Kolata in the *New York Times*, 3 March 1997.

p. 100   "Sheep Cloned by Nuclear Transfer from a Cultured Cell Line": K. H. S. Campbell, J. McWhir, W. A. Ritchie, and I. Wilmut, *Nature* 380 (1996): 64–66.

p. 100   Dolly's existence was announced to the scientific community: I. Wilmut, A. E. Schnieke, J. McWhir, A. J. Kind, and K. H. S. Campbell, "Viable Offspring Derived from Fetal and Adult Mammalian Cells," *Nature* 385 (1997): 810–13.

9   HUMAN "CUTTINGS"

p. 103   In fact, there is no scientific basis: As I describe in the text, the major genetic risks associated with natural reproduction are likely to be reduced for children born through cloning. However, various scientists have suggested that other genetic problems—unique to the cloning process—could arise. Three specific concerns have been mentioned.

The first concern focuses on specialized DNA structures called *telomeres* that act as protective sealing agents at both ends of every chromosome. During the normal process of DNA copying that occurs with each cell duplication, the telomeres get whittled down slightly. Some scientists believe that the gradual loss of telomeres over a lifetime is a major—perhaps even the primary—contributing factor to aging. But obviously, there must be a way to build telomeres back up in each new person and animal, or life on earth would go extinct. A special cellular enzyme called *telomerase* does just that, but it appears to be present only in germ cells and the early embryo. So the worry is that the donor cell used for cloning will have "aged" chromosomes that will impart a shorter life span, and perhaps increased disease susceptibility, to the child that is born.

Although this possibility can't be ruled out until animal experimentation is completed, it seems unlikely owing to a fundamental principle of developmental biology—feedback compensation. An embryo can break into four separate parts—each a quarter the size of the original—and yet the identical quadruplets that emerge are all full-size children. Similarly, it seems likely that the early embryo will compensate for telomere lengths that are initially shorter than normal by working harder to bring them up to full size.

The second concern focuses on a process known as *genomic imprinting*, which refers to the finding (in the last fifteen years) that certain genes function differently depending on whether they are received from the mother or the father. These functional differences are maintained by chemical modifications of the DNA. There is some concern that the genomic imprinting of DNA in donor cells used for cloning will be different from the imprinting found in a normal one-cell embryo, and that such a difference could affect development. Although, once again, we won't know for sure if this is a problem until experiments in animals are completed, the very fact that Dolly turned out to be a healthy lamb suggests otherwise (at least for sheep). If healthy monkeys can be born through cloning (from adult cells), then genomic imprinting in humans won't be a problem either.

The final concern focuses on *chromosome remodeling*. This is not a genetic concern per se, but a biochemical one. In order for cloning to work, the protein signals present on the donor cell DNA must be replaced by protein signals provided by the egg. If all the protein replacements are not made precisely as required, the correct developmental program of gene activity will not be carried out. Incomplete chromosome remodeling is probably responsible—to a large extent—for there being only one sheep born out of the 277 embryo-donor cell fusions accomplished by Ian Wilmut.

In the end, chromosome remodeling is likely to represent the predominant technical concern in a consideration of the feasibility and safety of cloning. Although I suspect that the success rate of proper chromosome remodeling will go up dramatically with further experimentation and optimization of conditions, a method for distinguishing between "properly remodeled" and "bad" embryos will probably be required to allow routine cloning to be performed on human beings.

p. 105   . . . cloning is "against God's will": Data extracted from a *Time*/CNN poll taken over 26 and 27 February 1997 and reported in *Time* on 10 March 1997, an ABC *Nightline* poll taken over the same period, and poll results reported in the *Chicago Tribune* on 2 March 1997.

p. 106   "It's a horrendous crime to make a Xerox (copy) of someone": Quoted in article by Jeffrey Kluger in *Time*, 10 March 1997.

p. 106   "Can the cloning create a soul?": Quoted in article by Carol McGraw and Susan Kelleher for the *Orange County Register*, 25 February 1997.

p. 106   "cloning would only produce humanoids": Quoted in the on-line version of the *Arlington Catholic Herald* (http://www.catholicherald.com/bissues.htm) on 16 May 1997.

p. 107   "synthetic humans would be easy prey": *New York Times* editorial, 28 February 1997.

p. 111   . . . a nationwide search conducted over a two-year period: Actually, one person was identified who showed tissue compatibility, but he backed out at the last minute, refusing to go through what is an intense and time-consuming protocol.

p. 111   Fourteen months later, the bone marrow transplant: This true life story was made into a television movie in 1993 called "For the Love of My Child: The Anissa Ayala Story."

p. 112   . . . instead of having genetic material that was 99.95 percent the same: Any two randomly chosen human beings are 99.9 percent genetically the same as each other, even if they have no ancestors in common for as far back as they can tell. If my genetic material is 99.9 percent the same as yours, then what do geneticists mean when they say that brothers and sisters share just 50 percent of their genes in common? The answer is that this 50 percent refers only to a *proportion of the genetic difference* that normally distinguishes two people from each other. In other words, I am 99.90 percent similar to an unrelated person, but 99.95 percent similar to my brother Bruce.

p. 112   "It's absolutely ethically wrong to have a child as a donor": Quoted in article by Craig Quintana in the *Orlando Sentinel Tribune,* 2 April 1990.

p. 112   "You are treating a human being as an object": Article by Mike Graham in *The Sunday Times* (London), 1 April 1990.

p. 112   "Children are not medicine for other people": Quoted in article by Michael Specter in the *Washington Post,* 25 March 1990.

p. 112   "What they're doing is ethically very troubling": Quoted in an article by David Gorgan, Nancy Matsumoto, and Kristina Johnson in *People,* 5 March 1990.

p. 112   "I can't think of a morally acceptable reason to clone a human being": Quoted in *Time,* 10 March 1997.

p. 112   Bioethicists and others who condemned the Ayalas: A number of the critics, including Arthur Caplan, admitted that the Ayalas were not being unethical based on their statement that they would love the child whether she was a match for Anissa or not. Instead, many suggested that what really worried them was the possibility that other families in need of a donor would have repeated abortions until a pregnancy was achieved with a compatible fetus. According to John Fletcher and others, this was clearly unethical. I must admit I don't understand the logic behind this kind of pronouncement from bioethicists who are otherwise pro-choice.

If one accepts a woman's right to have an abortion for any reason that she chooses, then it seems that donor incompatibility is just as good a reason as "I don't feel like having a baby at the moment."

p. 112   *Can anybody out there provide a universal definition*: The *Washington Post*, 25 March 1990.

p. 113   . . . millions of people still have babies without any forethought at all: As late as 1987, the Alan Guttmacher Institute estimated that 54 percent of all pregnancies in the United States were unintended, as quoted in *The Record* on 15 April 1990.

p. 116   The child that is born would be related by genes to one mother: The woman whose genetic material was used to create the child is not the genetic mother, as we will discuss more fully below. But since she will be raising the child in the context of a family, it is still appropriate to call her a mother.

p. 117   How many people would actually want to clone themselves?: The same percentage was obtained in a *Time*/CNN poll taken over 26 and 27 February 1997 and reported in *Time* magazine on 10 March 1997, and an ABC *Nightline* poll taken over the same period, with results reported in the *Chicago Tribune* on 2 March 1997. The ABC poll question was "If it becomes possible, would you personally like to be cloned—to have a child who would look exactly like you and have an exact copy of your own genes?"

p. 120   . . . "engineering someone's entire genetic makeup" Op-ed piece by Daniel Callahan in the *New York Times*, 26 February 1997.

p. 122   . . . "cloning's identicality would restrict evolution": Column by William Safire in the *New York Times*, 27 February 1997.

p. 122   "What nonsense, what utter utter nonsense": Speaking at a Senate committee hearing on cloning as quoted in an article by Gina Kolata in the *New York Times*, 13 March 1997.

p. 123   Bahamas-based company called Clonaid: The *Business Wire*, 10 March 1997.

p. 123   For in the end, international borders can do little: On March 6, 1997, the Roslin Institute, home of Dolly, was granted two international patents for the "cloning of animals," by the World Intellectual Property Organization (WIPO), a United Nations agency based in Geneva, Switzerland. The patent application was purposely worded to be inclusive of human cloning so that the inventors could use it as a legal vehicle to try to prevent this particular application from being used by anyone else (since the Roslin scientists are strongly opposed to it). It is doubtful, however, that the fear of patent infringement will have any effect on cloning enterprises that operate in countries that refuse to accept the WIPO ruling. And in any case, the patent expires in 2017.

10   WHERE WILL CLONING LEAD US?

p. 127   Skin cells can't be converted into bone marrow cells: At least not yet! We also thought until recently that none of these adult cells could be converted back into an embryo.

p. 128   . . . embryologists in the United States and England: G. R. Martin, "Isolation of a Pluripotent Cell Line from Early Mouse Embryos Cultured in Medium Conditioned by Teratocarcinoma Stem Cells," *Proceedings of the National Academy of Sciences U.S.A.* 78 (1981): 7634–38; M. J. Evans and M. H. Kaufman, "Establishment in Culture of Pluripotential Cells from Mouse Embryos," *Nature* 292 (1981): 154–56.

11   THREE MOTHERS AND TWO FATHERS

p. 134   a child can now have two bio-moms: With cloning, the number of possible bio-moms increases to three. In addition to the gene-mom and birth-mom, a third woman can contribute the egg cytoplasm. The egg cytoplasm carries organelles called mitochondria that contain their own DNA, which is copied and transmitted to every cell in the body. The amount of genetic information carried by mitochondria is 200 million times less than that carried in the nucleus of a human cell, and it typically makes no contribution to inherited differences among individual people. However, there are some very rare diseases caused by mutations in mitochondrial DNA. A woman who carries such a disease could pass it on through an unfertilized egg used in the process of cloning (with a nucleus obtained from another woman).

p. 135   *Having a Baby Without a Man*: Susan Robinson and H. F. Pizer (New York: Simon & Schuster, 1985).

p. 135   If she desires an anonymous source: Sperm banks accredited by The American Association of Tissue Banks are listed at the following web site: http://www.fertilitext.org/banks.html.

p. 136   up to 6 million children may be living in such families: Carole Cullum, "Co-Parent Adoptions: Lesbian and Gay Parenting," *Trial* 29 (1993): 28.

p. 136   an egg, a sperm nucleus, and a womb: My inclusion of a *womb*, rather than a *woman*, as a biological ingredient, is not meant to dehumanize women or to suggest that they are simply babymaking machines. Rather, it is to keep the list of ingredients as specific as possible. A womb will always be required for fetal development, but it need not be a uterus, nor even inside a woman, as we will discuss in chapter 16.

p. 137 . . . 2 million infertile American couples: Martha A. Field, *Surrogate Motherhood* (Cambridge, Mass.: Harvard University Press, 1988), p. 162.

12 CONTRACTING FOR A BIOLOGICAL MOTHER

p. 139 A 1978 *Time* magazine article: "The Cloning Era Is Almost Here," 19 June 1978, p. 100.

p. 139 . . . the term *surrogate mother*: The definition of a "surrogate mother" in the *Oxford English Dictionary* (OED) is "A woman whose pregnancy arises from the implantation in her womb of a fertilized egg or embryo from another woman." But this definition is followed by the contradictory example of a surrogate who conceives through the process of artificial insemination. The OED *definition* describes what we now refer to as gestational surrogacy, while the OED *example* is one of traditional surrogacy. A better all-encompassing definition of surrogate motherhood is provided by *The Encyclopaedia Britannica* as "a practice in which a woman (the surrogate mother) bears a child for a couple unable to produce children in the usual way, usually because the wife is infertile or otherwise unable to undergo pregnancy."

p. 139 "she is the *actual* mother": Martha A. Field, *Surrogate Motherhood* (Cambridge, Mass.: Harvard University Press, 1988), p. 5.

p. 139 the pregnant woman is indeed a biological surrogate: Strictly speaking, the surrogate mother should be considered only a *prospective* surrogate during the time she is pregnant. Her full status as a surrogate mother would only be conferred after her pregnancy is completed and she has relinqushed the baby.

p. 140 they will have to turn to a commercial surrogacy agency: A recent alternative is the use of classified advertisements posted by women who wish to act as surrogates and others who desire the services of a surrogate. The American Surrogacy Center, Inc. maintains lists of both types—with messages, personal descriptions, and sometimes photos—at its World Wide Web site (http://www.surrogacy.com). Working directly with a prospective surrogate rather than going through an agency entails certain risks that will be discussed shortly.

p. 140 Total costs can run up to $50,000: This figure is based on information provided by the Center for Surrogate Parenting & Egg Donation at its web site (http://www.surroparenting.com/surrpar.html).

p. 141 when, for example, a mother or sister acts as a surrogate mother: Leon Kass, "The Meaning of Life—In the Laboratory," in *Ethics,* ed. Alpern, p. 98–116.

p. 142 "What is fundamentally unethical about surrogate mother arrangements": Herbert T. Krimmel, "Surrogate Mother Arrangements from the Perspective of the Child," in *The Ethics of Reproductive Technology,* ed. Kenneth D. Alpern (New York: Oxford University Press, 1992), pp. 57–70.

p. 142 . . . "feel no less bonded to their children than responsible genetic parents": Margaret Jane Radin, "Market-Inalienability," in Ibid., pp. 174–94.

p. 142 a few prominent ethicist-lawyers, such as John Robertson: John Robertson, *Children of Choice: Freedom and the New Reproductive Technologies* (Princeton, N. J.: Princeton University Press, 1994), pp. 130–32.

p. 142 "Symbolic arguments and pejorative language seem to make up the bulk": Lori Andrews, "Surrogate Motherhood: The Challenge for Feminists," in *Ethics,* pp. 205–19; reprinted from Lori B. Andrews, "Surrogate Motherhood: The Challenge for Feminists," *Law, Medicine and Health Care* 16 (1988): 72–80.

p. 142 "I find it extremely insulting that there are people saying": Ibid.

p. 144 . . . time commitment and discomfort experienced by an egg donor: As pointed out by various critics, the use of the term "donor" to describe the individual who provides sperm or eggs is not strictly appropriate when the donor is paid for his or her donation. Nevertheless, like so many other not-strictly-appropriate reprogenetic phrases, this terminology has become widely accepted and used by the public and the media. As such, I will continue to use it throughout this book.

p. 146 "Seeing her, holding her, she was my child": Field, *Surrogate,* p. 3.

p. 148 In addition to New Jersey, a number of states: Legality can be determined in two ways—through court decisions in response to lawsuits brought when surrogacy disputes arise (case law), or through statutes passed by state legislatures (statutory law). The legal status of surrogacy in any state is always subject to change with new legislation or new court decisions. In many states not mentioned here, the prevailing law is still in flux at the time of this writing. Those interested in the current status of surrogacy law in any state should consult the World Wide Web site maintained by The American Surrogacy Center, Inc. and their coded map of the United States at http://www.surrogacy.com/legals/map.html.

p. 151   *No surrogate should be allowed to participate in a program*: Quoted from material posted at the web site maintained by The American Surrogacy Center (http://www.surrogacy.com/agencies/articles/litz/index.html).

13   BUYING AND SELLING SPERM AND EGGS

p. 152   artificial insemination by donor: Until the mid-1980s, the term "Artificial Insemination by Donor" with the acronym AID was more commonly used. But the similarity of this acronym to AIDS (which, by chance, also has a connection to sperm and semen) can cause confusion. For this reason, the use of the term "Donor Insemination," with its acronym DI, has become more common.

p. 152   The Italian priest and physiologist: John Timson, "Lazzaro Spallanzani's seminal discovery," *New Scientist*, 13 December 1979; K. J. Betteridge, "An Historical Look at Embryo Transfer," *Journal of Reproduction and Fertility* 62 (1981): 3.

p. 152   And in the 1790s, the Scottish physician: Gina Maranto, *Quest for Perfection: The Drive to Breed Better Human Beings* (New York: Scribner, 1996), p. 132.

p. 153   . . . "the handsomest student in his class": Quoted in Robert Francoeur, *Utopian Motherhood: New Trends in Human Reproduction* (London: George Allen & Unwin, 1971) pp. 11–13.

p. 153   Some legislatures and courts: Gena Corea, "The Subversive Sperm: A False Strain of Blood," in *Ethical Issues in the New Reproductive Technologies,* ed R. Hull (Belmont, Calif.: Wadsworth, 1990), pp. 56–68; Maranto, *Quest,* pp. 169–72.

p. 154   *The donor may not be feeling particularly well*: R. Snowden and G. D. Mitchell, *The Artificial Family: A Consideration of Artificial Insemination by Donor* (London: George Allen & Unwin, 1981), pp. 68–69.

p. 154   . . . almost always excluded unmarried women and lesbians: Office of Technology Assessment, U.S. Congress, *Artificial Insemination: Practice in the United States: Summary of a 1987 Survey—Background Paper,* OTA-13P-BA-48. (Washington, D.C.: United States Government Printing Office, 1988).

p. 155   And beginning in 1978, a series of women-friendly health centers: Ibid. The first of these were the Vermont Women's Health Center and the Oakland Feminist Women's Health Center in Oakland, California, which has its own sperm bank. A comprehensive list of fertility doctors and sperm banks that are friendly to single women and lesbians is provided at the "Lesbian Mom's Web Page": http://www.lesbian.org/moms/drs_sb.htm.

p. 155 A comprehensive 1987 U. S. government survey: Office of Technology Assessment, *Artificial Insemination.*

p. 156 . . . the Center for Surrogate Parenting & Egg Donation: See http://www.surroparenting.com.

p. 156 What this fertility center and others: See for example The Atlanta Reproductive Health Centre at the World Wide Web address: httpp://www.ivf.com/dnr2col.html.

p. 157 Thomas M. Pinkerton, writing for The American Surrogacy Center: See http://www.surrogacy.com.

p. 158 . . . the donor was a medical student: My Princeton colleague, and historian of science, Professor Angela Creager has pointed out to me that, whether consciously or subconsciously, physicians who controlled DI over the decades have attempted to *reproduce themselves.* She suggests that this is "another irony of convoluted reproduction woven into the complications of reprogenetic techologies."

p. 158 . . . many practitioners selected only those who appeared to possess: R. Snowden and G. D. Mitchell, p. 64.

p. 159 . . . shift the distribution of genes: No currently available screening procedure can completely eliminate the risk of genetic disease in a child conceived with donor sperm. Screening procedures will miss latent mental and physical ailments that do not appear until much later in a donor's life. In addition, donors can be carriers of hidden disease genes that are also carried, unknowingly, in the recipient woman's genetic material so that a DI child could express the disease. Nevertheless, in *statistical* terms, effective donor screening and selection reduce the likelihood of genetic disease relative to the population at large.

p. 160 "If intellectual qualities were inheritable": Editorial page, *New York Times,* 27 May 1982.

p. 160 "This is a gimmick, an unrealistic hope for families": *United Press International,* 12 June 1982.

p. 162 comments made in 1994 by several satisfied parents: All of the quotes were derived from a story by Jennifer Bojorquez, "In His Image," *Sacramento Bee,* 19 December 1994.

14 CONFUSED HERITAGE

p. 163 . . . "one's own child": This phrase is commonly used and understood to mean a child conceived with one's own gamete, either sperm or egg.

However, the use of the phrase in this exclusive way is demeaning to the strong parent-child relationship that can exist between adopted children and their adoptive social parents. For this reason, I have avoided its use wherever possible. At this point, however, I have chosen to use the phrase as a setup to challenge its meaning, as will become clear shortly.

p. 163   Adoption of unrelated children was extremely rare: I refer here to adoption in the modern Western sense of the term. According to the *Encyclopaedia Britannica,* "In most ancient civilizations and in certain later cultures as well, the purposes served by adoption differed substantially from those emphasized in modern times. . . . The person adopted invariably was male and often adult. In addition, the welfare of the adopter in this world and the next was the primary concern; little attention was paid to the welfare of the one adopted."

p. 166   the particular DNA molecules present in a human egg: In the fertilized egg that eventually develops into a child, a mother deposits just a single copy of DNA for each of the twenty-three human chromosomes. A second set of twenty-three DNA molecules is deposited in this same egg by the genetic father. The information present in each of these forty-six DNA molecules is then copied over time into 100 million million *new* sets of DNA molecules that are placed into each new cell formed during fetal and child development. Each of these new DNA molecules is built from raw materials that are recovered from the food that the mother, and then the child, consumes.

   Where do the twenty-three DNA molecules that actually come from the mother end up? Well, most—if not all—of them disappear long before the child is born. Fewer than one out of eight cells in the early embryo actually end up in the fetus. The remaining cells—with at least 87 percent of the original parental DNA—are channeled into the placenta or uterine linings, which are ejected from the mother's body and discarded as medical waste after birth. Of the motherly DNA molecules that survive in the fetus itself, many end up in short-lived cells such as those in the blood, the skin, or the intestines, which constantly degenerate to be replaced by newly made cells. When cells die or are discarded, the DNA molecules within them disintegrate into the small molecules, or single atoms, from which they were originally built. Thus, at most, only a handful of original DNA molecules from the mother survive in a few scattered cells among the 100 million million present in the child's body.

p. 167   . . . Tim was born without testicles: The medical term *anorchia* is used to describe the condition of a boy who is born without testicles but with a penis. Fetal development of a penis can only occur in the presence of testicular tissue. Thus, immature testes must have been present in the developing fetus, with degeneration occurring for unknown reasons prior

to birth. In the particular case of Tim Twomey, it is clear that degeneration had to be caused by nongenetic factors since the same medical
condition did *not* appear in his identical twin brother.

p. 167 . . . Dr. Silber performed the transplantation: Sherman J. Silber, "Transplantation of a Human Testis for Anorchia," *Fertility and Sterility* 30
(1978): 181–87.

p. 167 . . . a 6-pound, 14-ounce baby boy: An interesting side note is that even
after Tim's sterility problem was cured, the Twomeys were still unable
to achieve pregnancy because of a subsequently discovered problem with
Jannie's menstrual cycle. This problem was eliminated with appropriate
medical treatment, and the Twomeys achieved pregnancy a few months
later. S. J. Silber and L. J. Rodriguez-Rigau, "Pregnancy after Testicular
Transplant: Importance of Treating the Couple," *Fertility and Sterility* 33
(1980): 454–55.

p. 168 . . . intellectualization conflicts with the primeval instinct: The philospher
Kenneth Alpern has described other interesting "genetic puzzles" that also
confuse the meaning of "one's own child." The most thought-provoking
of these is one in which a woman walking down the street happens to
discover a baby in a stroller with a genetic makeup that is identical to
her own, just by chance. In a variation of this scenario, one can imagine
that the baby actually shares only half of its genetic material with the
woman walking down the street, so that it would appear—by all imaginable tests—to be that woman's child. Alpern asks whether the woman
should view this child as "her own," even if she has no reproductive link
to it. He concludes, "The science of genetics certainly does not provide
full answers to the questions that we have been asking . . ." In fact,
Alpern is wrong in his conclusion because of a failure to appreciate the
distinction between the *ends* (increased transmission of genes) and *means*
(the instinctive desire to have children that are physically connected) that
operated during the process of evolution. Kenneth D. Alpern, "Genetic
Puzzles and Stork Stories: On the Meaning and Significance of Having
Children," in *The Ethics of Reproductive Technology*, ed. Kenneth D. Alpern
(New York: Oxford University Press, 1992), pp. 147–69.

p. 168 "would radically alter the very definition of a human being": Quoted in
article by Earl Lane in *Newsday*, 13 March 1997.

p. 171 Since 1991, reprogeneticists have been able to extract: K. Y. Cha, J. J.
Koo, J. J. Ko, D. H. Choi, S. Y. Han, and T. K. Yoon, "Pregnancy after
In Vitro Fertilization of Human Follicular Oocytes Collected from Non-
stimulated Cycles, Their Culture In Vitro and Their Transfer in a Donor
Oocyte Program," *Fertility and Sterility* 55 (1991): 109–13.

p. 171 "The idea is so grotesque as to be unbelievable": Quoted in article by Gina Kolata in the *New York Times,* 6 January 1994.

p. 171 "It would be devastating to grow up knowing": Ibid.

p. 172 "Over all, you avoid more pain and suffering": Ibid.

p. 173 . . . transplanted rat spermatogonia into mouse testes: D. E. Clouthier, M. R. Avarbock, S. D. Maika, R. E. Hammer, and R. L. Brinster, "Rat Spermatogenesis in Mouse Testis," *Nature* 381 (1996): 418–21.

p. 174 . . . spermatogonial cells are easily frozen: M. R. Avarbock, C. J. Brinster, and R. L. Brinster, "Reconstitution of Spermatogenesis from Frozen Spermatogonial Stem Cells," *Nature Medicine* 2 (1996): 693–96.

p. 174 "Part of the way we think of who we are": Quoted in article by Gina Kolata in the *New York Times,* 30 May 1996.

15   SHARED GENETIC MOTHERHOOD

p. 176 A certain type of happily bonded couple: Although at the time of this writing, there were no publicized cases of shared *genetic* motherhood, there was at least one attempt at shared *biological* motherhood between members of a same-sex couple. On August 25, the *Mail On Sunday* (a British tabloid) reported that a lesbian couple had asked an IVF practitioner to retrieve eggs from one of them, fertilize the eggs with donor sperm, and then introduce them into the uterus of the second woman. The resulting baby would then be raised by two biological mothers— one would be her gene-mom, the other her birth-mom—who would "share in the experience of motherhood." Unfortunately for this couple, the physician took their request to his hospital's ethics review board, which ruled against it. Although this couple failed in its attempt to reach its reproductive goal, it seems likely that others have pursued the same goal with success, away from the eyes of the press, and close-minded male medical personnel.

p. 177 But rather than discarding this pronucleus: To allow the text to flow better, I have skipped some details here. The maternal pronucleus is actually recovered from the second egg with a bit of cytoplasm and membrane around it. This enclosed pronucleus is equivalent to a minicell. It is placed into the space between the membrane of the first egg and its zona pellucida coat. A chemical or electrical stimulant is then used to fuse the large egg and the minicell, bringing the foreign pronucleus into the new egg cytoplasm.

p. 178   This exact experiment was actually performed: J. McGrath and D. Solter, "Completion of Mouse Embryogenesis Requires Both the Maternal and Paternal Genomes," *Cell* 37 (1984): 179–83.

p. 178   . . . it couldn't develop properly into a live-born animal: This finding, since confirmed by other scientists, contradicts a claim made in 1976 by a Swiss embryologist named Karl Illmense for the production of not only live-born double-mothered mice, but also live-born mice who were derived solely from a single female parent. In the early 1980s, it was discovered that Illmense had faked his results, and that he never derived mice the way he claimed he had.

p. 179   . . . a different approach to producing mice: A. K. Tarkowski, "Mouse Chimaeras Developed from Fused Eggs," *Nature* 190 (1961): 875–60.

p. 179   Tarkowski's simple method: A cookbook-like protocol for the production of mouse chimeras is described in B. Hogan, R. Beddington, F. Costantini, and F. Lacy, *Manipulating the Mouse Embryo: A Laboratory Manual, Second Edition* (Cold Spring Harbor, NY: Cold Spring Harbor Press, 1994). An incredible picture of a chimera between a goat and sheep is shown on page 37 in C. R. Austin and R. V. Short, *Reproduction in Mammals; Book 5 Manipulating Reproduction* (Cambridge, Eng.: Cambridge University Press, 1986), and on the cover of the 16 February 1984 issue of *Nature*.

p. 179   Embryos and animals formed by combining cells: Some investigators object to the appellation chimera because of its negative connotations. They prefer the use of terms like "tetraparental" to describe, for example, an animal formed by mixing together two embryos with different mothers and fathers. However, chimera is still the term used by most animal embryologists.

p. 180   . . . more than one hundred natural-born chimeric human beings: G. Krob, A. Braun, and U. Kuhnle, "True Hermaphroditism: Geographical Distribution, Clinical Findings, Chromosomes and Gonadal Histology," *European Journal of Pediatrics* 153 (1994): 2–10; Patricia Tippett, "Human Chimeras," in *Chimeras in Developmental Biology* (London: Academic Press, 1984); A. J. Green, D. E. Barton, P. Jenks, J. Pearson, and J. R. Yates, "Chimaerism Shown by Cytogenetics and DNA Polymorphism Analysis," *Journal of Medical Genetics* 31 (1994): 816–17; S. Uehara, M. Nata, M. Nagae, K. Sagisaka, K. Okamura, and A. Yajima, "Molecular Biologic Analyses of Tetragametic Chimerism in a True Hermaphrodite with 46,XX/46,XY," *Fertility and Sterility* 63 (1995): 189–92.

p. 181   . . . the same thing should happen spontaneously: You may wonder why chimerism doesn't happen more often. It's because fertilized eggs are typically enshrouded within a Teflonlike zona pellucida coat until right before they implant into the uterine wall. The zona coat protects the integrity of each individual embryo and prevents the embryos themselves from coming into contact with each other. Thus, chimera formation by embryo fusion could occur in only two ways. Either both fertilized eggs lose their zona coat prematurely and then bump into and stick to each other, or they both lose their zona coat at the appropriate time but implant directly adjacent to each other, merging into one as they develop. The latter process seems likely to be responsible for the formation of most naturally born chimeras.

p. 181 . . . the tissues that differentiate into the sex organs: Male and female sex organs differentiate out of the same sexless tissues that appear early during fetal development. The direction of development is determined by the presence or absence of a signal transmitted by a gene on the Y chromosome. If the Y signal is made, it counteracts the natural tendency of the fetus to develop into a female. Fetal gonads become ovaries if the Y chromosome is absent, testicles if a Y is present. The fetal phallus develops into a clitoris or a penis, and the external tissue below it develops into a scrotum or the outer folds of the vulva.

p. 181 But the gonads themselves: A person is considered a true hermaphrodite when he/she has both ovarian and testicular tissue (irrespective of genital appearance). A person is considered intersex when the external genitalia show a mixture of male and female characteristics. Hermaphroditism and intersexuality can occur together or separately. About 25 percent of human hermaphrodites are chimeric. A variety of other developmental or genetic defects are responsible for other cases of intersexuality and true hermaphroditism. G. Krob, A. Braun, and U. Kuhnle, "True Hermaphroditism."

p. 181 . . . many intersex chimeras who have developed: Aside from the possibility of a patchy complexion or hair color, same-sex chimeras do not appear to be susceptible to any other physiological or behavioral abnormality. In fact, based on our understanding of development, there is no reason why they should be. All of the cells in a chimeric embryo are human. They all have the ability to produce and respond to the same range of molecular signals. And each cell—no matter what its origin—is programmed to work together with its neighbors to differentiate into each of the tissues in the adult human body. If there were one word to describe how cells behave in the developing human embryo and fetus, it would be *teamwork*. And with teamwork in action between even genetically distinct cells, every organ in the adult body can be built to function as it should.

p. 181 . . . 50 percent would be intersex individuals: This percentage is based on putting together two embryos without any knowledge of their sex. In this case, the probability that the first embryo is male is 0.5 and the probability that the second embryo is male is 0.5, giving a combined probability of 0.5 × 0.5 = 0.25 of having an all-male chimeric embryo. Similarly, the probability of an all-female chimeric embryo is also 0.25. The remaining 50 percent of embryos must therefore be intersex.

p. 184 . . . a machine called a flow cytometer: G. Levinson, K. Keyvanfar, J. C. Wu, E. F. Fugger, R. A. Fields, G. L. Harton, F. T. Palmer, M. E. Sisson, K. M. Starr, L. Dennison-Lagos, et al., "DNA-Based X-enriched Sperm Separation as an Adjunct to Preimplantation Genetic Testing for the Prevention of X-Linked Disease," *Human Reproduction* 10 (1995): 979–82.

p. 188 Yet in terms of general health characteristics: In fact, chimeric children may have *greater* disease resistance because they carry four different copies (rather than two) of each immune function gene. A larger number of immune function genes can provide an individual with a greater chance of resistance to any random infectious disease.

16   COULD A FATHER BE A MOTHER?

p. 194 If a woman's abdomen can act as womb: A physician and researcher named Cecil Jacobson claimed to have initiated and maintained a pregnancy in the abdominal cavity of a male baboon during the mid-1960s at George Washington University Medical School. ("Not Half-Dad," story by Julie Wheelwright in the *Guardian,* 9 April 1992) Although Jacobson never published this work, various scientists and writers have used his claim as support for the contention that male pregnancy could be achieved in humans as well. However, in 1992, the same Cecil Jacobson was convicted on fifty-two counts of fraud and perjury for using his own sperm surreptitiously in the artificial insemination of infertile women and for falsely claiming that women treated in his clinic had become pregnant when they had not. This latter turn of events clearly invokes skepticism over the validity of his earlier claims.

p. 194 *"Abdominal pregnancy is a rare but life-threatening condition"*: A. Wagner and A. J. Burchardt, "MR Imaging in Advanced Abdominal Pregnancy. A Case Report of Fetal Death," *Acta Radiol* 36 (1995): 193–95.

p. 194 *"Morbidity and mortality for both the fetus and the mother are considerable"*: W. A. Alto, "Abdominal Pregnancy," *American Family Physician* 41 (1990): 209–14.

p. 194 *"Care of the patient afflicted with it may present formidable challenges"*: S. Yu, J. A. Pennisi, M. Moukhtar, and E. A. Friedman, "Placental Abruption in Association with Advanced Abdominal Pregnancy. A Case Report," *Journal of Reproductive Medicine* 40 (1995): 731–35.

p. 195 These will include some transsexuals: According to a representative from the Transsexual Support Group UK, "The majority of transsexuals would like to reproduce. Yet there is little or no research being done in America primarily because of the moral issues involved." Wheelwright, "Not Half-Dad," *Guardian*.

17 THE VIRTUAL CHILD

p. 201 . . . cognitive starting points that were below average: Our genes provide each of us with a human mind that contains within it the capacity to think and reflect in ways that go far beyond that possible for even our nearest evolutionary cousin, the chimpanzee. No modern-day biologist debates the veracity of this statement. What has been debated for many years, however, is the contribution that genetic differences among people make to individual human differences in cognitive processes, personality, and other aspects of behavior. It is now clear that genes and environment work together in the expression of nearly every single human trait. (For an excellent discussion of the large body of scientific data that bear on this question, see Robert Plomin, *Nature and Nurture: An Introduction to Human Behavioral Genetics* [Pacific Grove, Calif.: Brooks/Cole Publishing, 1990].)

Which factor—genes or environment—is more important in the expression of a particular trait will often come down to the particular circumstances in which an individual person is placed by the vagaries of life. It is for this reason that percentage values calculated by scientists for the contribution of each cannot be applied to individual human beings. Professional geneticists who work in the field are well aware of this limitation. In fact, the basic term under investigation—genetic contribution—is defined *only* in the context of the particular population analyzed, and not for individuals within that population. In the analysis of a different population, a different genetic contribution might be determined. Unfortunately, these distinctions are often not appreciated by the public at large.

The best way to view the genetic contribution to personality and cognitive abilities is in the context of a set of *starting points*. Every person is born with hundreds of starting points for every separable aspect of who he or she is. But, as the psychology writer Winifred Gallagher says in her comprehensive account of the contribution of both nature and nurture to a person's identity or *I.D.*, "We're limited less by our vast genetic potential than by the narrow use that most of us and our environments make of it." (Winifred Gallagher, *I.D.: How Heredity and Experience Make You Who You Are* [New York: Random House, 1996].)

A prime example can be seen in the deplorable level of mathematics that is taught to American children in many public elementary schools, especially within urban areas. Most children have the innate capacity to achieve much more scholastically than they are ever given the opportunity to do, for a host of environmental reasons.

On the other hand, there are realms of intellectual, artistic, and physical prowess that most of us will never reach no matter how hard we might try. We are all aware of these limitations, and we are all aware that others can excel in some areas that we cannot. It is in this very personal way that each of us can recognize the contribution of genes to an individual's abilities.

p. 202   ... a tendency toward long-term happiness: Each of the particular aspects of temperament listed here are known to have a high degree of genetic influence. For an overview see Plomin, *Nature and Nurture.* In 1996, a series of papers demonstrated, for the first time, links between specific genes and particular personality components. The results obtained for risk-taking behavior were reported by J. Benjamin, L. Li, C. Patterson, B. D. Greenberg, D. L. Murphy, and D. H. Hamer, "Population and Familial Association Between the D4 Dopamine Receptor Gene and Measures of Novelty Seeking," *Nature Genetics* 12 (1996): 81–84; and R. P. Ebstein, O. Novick, R. Umansky, B. Priel, Y. Osher, D. Blaine, E. R. Bennett, L. Nemanov, M. Katz, and R. H. Belmaker, "Dopamine D4 Receptor (D4DR) Exon III Polymorphism Associated with the Human Personality Trait of Novelty Seeking," *Nature Genetics* 12 (1996): 78–80.

The results obtained for anxiety were reported by K. P. Lesch, D. Bengel, A. Heils, S. Z. Sabol, B. D. Greenberg, S. Petri, J. Benjamin, C. R. Muller, D. H. Hamer, and D. L. Murphy, "Association of Anxiety-Related Traits with a Polymorphism in the Serotonin Transporter Gene Regulatory Region," *Science* 274 (1996): 1527–30.

p. 203   two pregnancies that had been established: A. H. Handyside, E. H. Kontogianni, K. Hardy, and R. M. L. Winston, "Pregnancies from Biopsied Human Preimplantation Embryos Sexed by Y-Specific DNA Amplification," *Nature* 344 (1990): 768–70.

p. 203   ... carriers of a serious disease mutation: Five couples took part in the first clinical trial. The female partner of each couple was a carrier for a mutation causing one of the following diseases—X-linked mental retardation, adrenoleukodystrophy, Lesch-Nyhan syndrome, or Duchenne muscular dystrophy (see Handyside, et al.). Each of these mutations occurs in a different gene located on the X chromosome. The male partner did not carry mutations at any of these genes.

All daughters receive one X chromosome from their mother and one X chromosome from their father. Therefore, even if they received a mutation from their mother, they would still receive a "good" copy of the gene from their father, and they would not express the disease. However, boys have only a single copy of the X chromosome, which they receive from their mother. If the single copy of a particular gene is mutant, a boy will express the disease.

p. 203 The total of all the information: Every species of animal, plant, and microbe is characterized by its own unique genome. Each species has its own unique set of genes that are spread out on a defined number of chromosomes.

p. 204 . . . sickle cell anemia, cystic fibrosis, Tay-Sachs, and Huntington Disease: A comprehensive genetic description of every inherited disease known to medical scientists is available at a World Wide Web site maintained by the Center for Medical Genetics, Johns Hopkins University in Baltimore and the National Center for Biotechnology Information, National Library of Medicine, Bethesda, Maryland. The name of this compilation is "Online Mendelian Inheritance in Man, OMIM." Not only does it contain a detailed synopsis of each disease, it also provides pointers to related informational resources including original publications, as well as DNA and polypeptide sequences. The OMIM web address is http://www3.ncbi.nlm.nih.gov/omim/. Each independent gene and disease in the database is assigned a number and a name. The gene mutated in the sickle cell anemia trait is called HEMOGLOBIN—BETA LOCUS. It is abbreviated with the symbol HBB and its assignment number is 141900.

p. 204 There is no chemical technique that can provide information: The modern era of molecular genetics and biotechnology began during the 1970s through the invention of three independent techniques. First, DNA cloning allowed the isolation of individual genes out of the human genome. Second, DNA sequencing provided a rapid means for reading the information contained within isolated genes. Third, DNA synthesis provided a means for creating novel DNA fragments that could be used for genetic engineering as well as to prime the further analysis of complex genomes and the genes they contain. Together, these techniques changed the face of biology and medicine by giving scientists the power to look into genomes with complete clarity, and the tools to understand how genes function.

But the science enabled by these techniques was not all-powerful. There were two main limitations. First, even after the sequence of a particular gene had already been determined, it could still take weeks of intensive work by a scientific team to clone and determine the sequences of other alleles at the same gene. The second limitation concerned the starting material for analysis. Although cloning could be accomplished with what was then considered to be small amounts of DNA, these small amounts still corresponded to the material present in thousands of cells.

Both of these limitations disappeared with the invention of PCR. If biotechnology was powerful before PCR, it became thousands of times more powerful after PCR.

p. 205   . . . the Polymerase Chain Reaction: The term is purposely analogous to "nuclear chain reaction." The chain reaction, in both terms, refers to the process by which an initial small event becomes amplified at each step in a "chain," with one product acting as a stimulus to induce the production of two further products, which act as stimuli to produce four products, and so on, leading to an exponential explosion—of DNA fragments with PCR, and of nuclear energy with a nuclear chain reaction.

p. 205   . . . a legend in what is still a very young field: A history of PCR, the people involved, and the climate in which it was invented is told in a book by Paul Rabinow, *Making PCR: A Story of Biotechnology* (Chicago: University of Chicago Press, 1996). For those with an interest in bringing the magic of PCR into the classroom, Cold Spring Harbor Laboratory has developed a videotape entitled "Introduction to a Decade of PCR," available from Cold Spring Harbor Press in Cold Spring Harbor, New York.

p. 205   *Mullis was being playful on an April evening in 1983*: This quote was obtained in November 1996 from the World Wide Web address http://www.uky.edu/~holler/mullis.html.

p. 207   . . . simple probability rules tell us: This extrapolation is based on standard methods of probability analysis. The probability of occurrence P of N independent events, each with a probability of p alone, is simply p raised to the power of N (p^N). If $p = 0.90$, then $P = (0.9)^{10} = 0.35$. This means, conversely, that there is a 65 percent chance that at least one of the ten genes tested will not yield a usable result.

p. 209   DNA chips that are set to revolutionize the practice of genetics: David Stipp, "Gene Chip Breakthrough," *Fortune,* 31 March 1997; Mark Chee, Robert Yang, Earl Hubbell, Anthony Berno, Xiaohua C. Huang, David Stern, Jim Winkler, David J. Lockhart, Macdonald S. Morris, and Stephen P. A. Fodor, "Accessing Genetic Information with High-Density DNA Arrays," *Science* 274 (1996): 610–14.

p. 209   . . . a detector for the presence or absence of a particular allele: Detection occurs through a process called *hybridization*. Hybridization allows molecular biologists to take advantage of the natural propensity of complementary strands of DNA to bind to each other. To understand hybridization, you must first understand the structure of the DNA molecule.

    The DNA molecule is composed of two very long chains that spiral around each other like a two-stranded rope—the classic double helix. Each strand is composed of a series of bases (A, C, G, or T). Contained within the sequence of bases is the genetic message. The second strand does not contain any genetic information that is not in the first. However, the second chain is *not* identical to the first; rather it is *complementary* to it, as I will explain in a moment.

The simplest way to think of the DNA molecule is as a long vertical ladder with evenly spaced flat horizontal steps. The two sides of the ladder represent the backbones of the two chains, and each step represents two bases with one jutting out from each backbone, meeting the other in the middle. The bases on a particular step are referred to as a "base pair." It is not possible for just any two bases to be paired on a single step; only complementary pairs are allowed. The bases G and C are complementary to each other, and the bases A and T are complementary to each other. Complementarity flows from the ability of bases to fit together physically, like two adjacent pieces in a jigsaw puzzle.

Let us say that the sequence along one chain of a DNA molecule is ATTGCG. This would imply a sequence for the second chain of TAACGC. And if the first chain was TAACGC, then the second chain would have to be ATTGCG. Like mirror images, complementary sequences reflect each other. This is why I can say they carry the *same* information even though they are not identical, just as a mirror image of a photograph contains the same information as the photograph itself.

The next important thing to know about DNA is that the chemical link that holds bases together in the middle of each step on the ladder is not the strong covalent bond responsible for keeping the three atoms of the water molecule H_2O together. Instead it is the weak hydrogen bond that makes different water molecules cling to each other in a raindrop (or in a straw while you are sucking a soft drink into your mouth from a glass). When the temperature is raised sufficiently, the two strands of the double helix fall apart from each other (just as a water drop will disperse into a vapor). When the temperature is lowered, complementary strands try to find each other to rebuild a double helix (just as dew drops form on leaves of grass in the coolness of a summer morning).

Each microscopic block on a DNA chip is covered with many identical copies of just one of the two DNA strands that represents a particular allele of a particular gene. The DNA sample to be tested is dissolved in a solution and heated so that its double helices all fall apart. Then the sample is placed on the chip, and the temperature is reduced. Now, if a strand in the sample is complementary to one on the chip, the sample strand will stick or *hybridize* to the chip at a particular block. After a sufficient time, the DNA that has not hybridized to one block or another on the chip is washed away, and the chip is analyzed automatically under a microscope to determine which blocks detected complementary strands (representing particular alleles) in the sample and which did not.

p. 209 . . . capacity doubling every eighteen months: Stipp, "Gene Chip Breakthrough."

p. 211 These women . . . want nothing more than to be mothers: Quoted in article by Barbara Stewart, "Tough Choices: In Vitro Vs. Adoption," *New York Times,* 8 January 1995.

p. 211 Do we want to have a society: Dean Hamer and Peter Copeland, *The Science of Desire: The Search for the Gay Gene and the Biology of Behavior* (New York: Simon & Schuster, 1994) p. 219.

p. 211 Why is it OK for people to choose the best house: Quoted in a story by Jennifer Bojorquez, "In His Image," *Sacramento Bee,* 19 December 1994.

p. 212 For some, the idea of a father choosing a genetic "gift" for his son is repellent: From a theater review by Alan F. Wright and A. Christopher Boyd, "Choosing Genes," *Nature* 383 (1996): 312. The play called *The Gift* was written by Nicola Baldwin.

p. 212 If procreative liberty gives women the right to abort: Bonnie Steinbock, "Ethical Issues in Human Embryo Research," in *Papers Commissioned for the NIH Human Embryo Research Panel, Volume II* (Bethesda, MD.: National Institutes of Health, 1994), p. 39.

p. 212 The real problem is not the one we most fear: Diane B. Paul, "Eugenic Anxieties, Social Realities, and Political Choices," in *Are Genes Us? The Social Consequences of the New Genetics,* ed. C. F. Cranor (New Brunswick, N.J.: Rutgers University Press, 1994), pp. 142–54.

p. 212 We mold and shape our children: Bojorquez, "In His Image."

p. 213 . . . the environment will play as critical a role: Modern-day biologists, especially molecular biologists, have often been accused by their colleagues in the social sciences of being too "gene-centric" in their view of life. "Life is much more than genes," they say, "and yet all you ever seem to study and experiment on is the gene!" In fact, our critics are right on both counts. Human life, in particular, is much more than genes, and yet the gene seems to be at the focus of most of the exciting new biomedical research that hits the front page of newspapers regularly.

In my opinion, the reason for this unbalanced state of affairs is not as sinister as it may seem. It is not because modern biologists think they can control the world if only they can control your genes. In truth, most basic science researchers have no such thoughts. Instead, what they care most about is solving the narrow biological problem that appears before them.

To solve that problem, a good scientist will use whatever tools she can get her hands on. And as we approach the end of the second millennium, the most powerful tools available for the analysis of many aspects of life and health are based on the use of genetics.

So the reason that so many biomedical scientists focus on genes is that it's easy (in the relative sense of the word). But contrary to what many social critics claim, a "genetic ideology" does not pervade the field of biomedicine. If an alternative set of tools (based on biochemistry or biophysics) superseded genetics in its power to dissect the fundamental processes of life, you can be sure that future biomedical scientists would rush to embrace it.

There is a truly unfortunate consequence of our focus on the gene. When the media report the results of exciting new studies that demonstrate a particular genetic contribution to a human trait or disease, they sometimes give the impression that the gene represents the *entire* contribution. Even when the best science reporters try to present a balanced picture, the public only remembers what *has* been discovered, not the large part that is still blank. Report after report of this kind can lead to a widespread sense of genetic determinism, even as geneticists themselves see the world otherwise in terms of statistics and individual probabilities.

p. 214   . . . the choice will have to be made: There is another option not mentioned here. After eliminating the embryos that carry the disease genotype, the physician could throw away the test results or mix all the remaining embryos together. Then he would have no way of distinguishing among embryos that carried the normal or carrier genotypes. Ask yourself whether it makes any sense to purposefully eliminate information that could benefit your child in some way.

p. 216   " 'Eugenics' is a word with nasty connotations": Paul, "Eugenic Anxieties," pp. 142–54.

p. 216   . . . eugenics referred to the idea that a society: A detailed history of eugenics is presented in Daniel J. Kevles, *In the Name of Eugenics: Genetics and the Uses of Human Heredity* (New York: Alfred A. Knopf, 1985).

p. 217   *The Quest for Perfection: The Drive to Breed Better Human Beings*: Gina Maranto (New York: Scribner, 1996).

p. 219   If the asteroid that hit our planet 60 million years ago: The asteroid crashed into the Yucatán peninsula of Mexico and caused the expulsion of so much dust into the sky around the world that the sun was blocked out for a period of years, leading to the death of plant life, which caused the death and extinction of all large animals, including the dinosaurs. Species of small animals, including rodentlike mammals, were able to survive within greatly reduced populations that scavenged the earth for seeds and other meager food substances. But when the dust settled, and the sun returned to the sky, plant life blossomed once again and the earth became a fertile place with wide open ecological niches. Mammals took advantage of this new situation and evolved explosively into the many species alive today, including human beings.

p. 219   Many prospective parents choose not to learn the sex of their child: Indeed, my wife and I chose this path for our first child, although not for our other two.

p. 220   If embryo selection were available to all people in the world: The probability that free worldwide access to embryo screening could become available is, as a matter of economics, lower than the probability that all poverty will be eliminated on a global scale. This was the political premise of Aldous Huxley's *Brave New World*, which, in 1997, seems more securely in the realm of fiction than any of the fantastical stories told in the epilogue to the book you are now reading.

p. 220   . . . *a hidden advantage to the gene pool*: The idea that a deleterious allele might cause *good* as well as bad is often misunderstood. The scientific basis for this principle is well illustrated by considering the mutation in the hemoglobin gene that causes sickle cell anemia (the SC allele). The SC allele is present at a frequency of 10 percent or higher in some African populations. This high frequency has come about because people who are simply SC carriers—with no disease symptoms—have resistance to malaria. In an environment where malaria is prevalent, SC carriers have a higher chance of surviving to adulthood and passing on the SC allele to their children. This advantage is counterbalanced by the birth of children afflicted with sickle cell disease (caused by two copies of the SC allele), which increases as the population frequency of the SC allele increases. Ultimately, the frequency of the allele reaches an equilibrium point.

    If we consider the individual, though, an important point is often missed: the SC allele provides no advantage whatsoever to a person who never comes into contact with the malarial parasite. This means that a normal SC carrier living in North America (where malarial infection is almost nonexistent) has no advantage over a person without the allele. On the contrary, the probability that an SC carrier will marry another SC carrier and have to deal with the potential birth of a diseased child is greater than the probability of becoming infected with the parasite.

    Mathematical models tell population geneticists that the mutant alleles responsible for a number of other genetic diseases that are frequent in people from particular ethnic groups or geographical regions—like cystic fibrosis, Tay-Sachs, and PKU—also *once* provided carriers with some advantage of some kind. But whatever the environmental factors were that allowed these mutations to persist, they are no longer present. Thus, there is no reason why anyone should *want* to be a carrier, and there is no reason for anyone to wish carrier status (for any one of the thousands of different disease alleles that have been identified) on some distant descendant.

p. 221   they are worried that the genetic elimination of mental illness: Gallagher, *I.D.*, pp. 38–39.

p. 221   . . . the use of hallucinogenic . . . drugs: My intent in making this provocative statement is *not* to promote the use of drugs for achieving altered states of mind, but rather to illustrate the absurd nature of the contention that mental illness can serve a useful purpose for society.

p. 222   Inoculation of children with the polio vaccine: I thank John Robertson (University of Texas Law School, Austin) for bringing this interesting comparison to my attention.

p. 222   "the genetic conditions the affluent are concerned to avoid": Philip Kitcher, *The Lives to Come* (New York: Simon & Schuster, 1996), p. 198.

p. 223   . . . such refusal would . . . force society to help the unfortunate children: Ibid, p. 201.

p. 223   "the notion that individual desires should sometimes be subordinated": Paul, *Eugenic Anxieties,* p. 148.

p. 224   . . . the technology will remain prohibitively expensive: Embryo screening will always be expensive because of the highly skilled laborers (physicians and Ph.D. scientists) who will always be required to carry it out. This means that it will never be available to those at the bottom end of the socioeconomic scale, although where the line will be drawn is hard to tell.

p. 225   . . . countries like Germany, Norway, Austria, and Switzerland: Lori B. Andrews and Nanette Elster, "Cross-cultural Analysis of Policies Regarding Embryo Research: Appendices," in *Papers Commissioned for the NIH Human Embryo Research Panel, Volume II* (Bethesda, MD.: National Institutes of Health, 1994) pp. 65–407. Current legislation to ban embryo selection is driven more by a religious desire to protect all embryos from harm than by a communitarian desire to equalize the goods and services available to all citizens.

18   THE DESIGNER CHILD

p. 227   "*At the crossroads where reproductive technology and genetic forecasting meet*": Margery Stein, "Making Babies or Playing God," *Family Circle,* 20 September 1994.

p. 228   This will always be possible whenever one parent is disease-free: A very rare situation in which this rule would fail to hold could occur if a person had two mutant copies of the Huntington Disease (HD) gene. All of the gametes produced by such a person would carry the disease mutation, and all embryos that emerged after fertilization would carry the mutation as well. Since HD is a dominant disease, this means that all embryos would have the disease genotype. The predicted frequency of this rare homozygous HD genotype in the world population is 1 in 100 million (but in isolated populations where the disease is prevalent, it could be higher).

p. 229 . . . "genetic engineering" and "gene therapy": The terms "gene therapy" and "genetic engineering" have been used to describe the alteration of genetic material in embryonic cells as well as cells in the body of an adult or child. These two different uses are distinguished by qualifiers: germ-line gene therapy is practiced on embryos; somatic cell gene therapy is practiced on cells obtained from a person's body. The big difference that many ethicists see between these two applications of the same basic technology is that the changes produced with somatic cell gene therapy cannot be passed on to future generations, while germ-line changes can. The logic used by many ethicists to support this point of view is that parents don't have the right to impose their genetic choices on their children and their children's children as well. According to this logic, the random inheritance of genes—even if they cause disease—is preferable to genetic choice—even if it is used to prevent disease.

Various applications of somatic cell gene therapy are already underway in clinical trials. However, I will not discuss somatic cell gene therapy in this book. Thus, I will use the simple terms "gene therapy" and "genetic engineering" to refer only to genetic changes made on embryonic cells that can be passed from one generation to the next.

p. 230 An alternative method was developed: J. W. Gordon, G. A. Scangos, D. J. Plotkin, J. A. Barbosa, and F. H. Ruddle, "Genetic Transmission of Mouse Embryos by Microinjection of Purified DNA," *Proceedings of the National Academy of Science USA 77* (1980): 7380–84. The method itself is simple in concept. First, a dish of fertilized eggs is placed within a special microscope that allows the clear visualization of the various internal embryonic components including the two pronuclei. While gazing through the microscope, a scientist uses one hand to control a special microscopic tool that allows her to pick up one embryo (at a time) and hold it securely in place. With her other hand, she controls a microscopic needle with a very sharp point that is brought up against the surface of the fertilized egg. The embryo is stabbed with the needle, whose tip is further positioned within one of the two pronuclei. Finally, the DNA molecules present within the needle are injected into the pronucleus, the needle is removed, and the embryo is released from its holder to fall back onto the bottom of the dish.

What happens next inside the little embryo is remarkable. The embryo actually *sees* the tiny DNA fragments that it has just received (in the way that any cell can see anything), and isn't happy that they're all alone. The experience of billions of years of evolution tells the cell that orphaned DNA fragments fall out of damaged chromosomes, and the best bet for survival is to stuff them back in somewhere. So that's just what the cell does with the injected foreign DNA. It randomly places the DNA back into a chromosome somewhere. And once the foreign DNA is spliced back into a standard chromosomal DNA molecule, its future is the same as the future of all the other genes in the embryo.

p. 230 *Manipulation of the Mouse Embryo*: B. Hogan, R. Beddington, F. Costantini, and E. Lacy, *Manipulating the Mouse Embryo: A Laboratory Manual* (Cold Spring Harbor, NY: Cold Spring Harbor Press, 1994).

p. 231 . . . 95 percent of the DNA in the genomes of all mammals: Genes occupy less than 3 percent of the three billion bases of DNA present in the human genome. There are other packets of genetic information that have nothing to do with gene activity but are useful to the cell in other ways. One type provides the DNA foundation for constructing a specialized chromosome structure (called a centromere) that acts as a physical handle for the movement of whole chromosomes between cells at the time of cell duplication and division. Another type provides protection for the sensitive ends of DNA molecules (telomeres). But even when all of the useful pieces of genetic information are added up, they account for just 5 percent of the genome.

What does the remaining 95 percent do? Probably nothing for you or me. The most likely explanation for its existence lies in the "selfish gene" concept promoted by Richard Dawkins in the book by the same name. According to Dawkins, all of evolution takes place at the level of individual DNA fragments and not at the level of persons, animals, or species. DNA fragments simply use us as a "survival machine," in the words of Dawkins. And while some DNA fragments are clearly needed to help the survival machine survive (these are our 100,000 genes), the rest of our DNA—in fact, the vast majority of it—is simply freeloading.

p. 231 . . . future reprogeneticists will gently put in whole chromosomes: Success has already been achieved with the placement of artificial human chromosomes into nonembryonic cells. See the news article by Nicholas Wade, "Artificial Human Chromosome Is New Tool for Gene Therapy," *New York Times*, 3 April 1997. The original publication is J. J. Harrington, B. Van Bokkelen, R. W. Mays, K. Gustashaw, and H. F. Willard, "Formation of De Novo Centromeres and Construction of First Generation Human Artificial Microchromosomes," *Nature Genetics* 15 (1997): 345–55.

p. 232 Gene replacement has already been achieved in animals: T. Doetschman, R. G. Gregg, N. Maeda, M. L. Hooper, D. W. Melton, S. Thompson, and O. Smithies, "Targeted Correction of a Mutant HPRT Gene in Mouse Embryonic Stem Cells," *Nature* 330 (1987): 576–78; K. R. Thomas and

M. R. Capecchi, "Site-Directed Mutagenesis by Gene Targeting in Mouse Embryo-Derived Stem Cells," *Cell* 51 (1987): 503–12; S. Thompson, A. R. Clarke, A. M. Pow, M. L. Hooper, and D. W. Melton, "Germ Line Transmission and Expression of a Corrected HPRT Gene Produced by Gene Targeting in Embryonic Stem Cells," *Cell* 56 (1989): 313–21.

pp.232– This new approach is called "anti-gene therapy": Larry A. Couture and
33   Dan T. Stinchcomb, "Anti-gene Therapy: The Use of Ribozymes to Inhibit Gene Function," *Trends in Genetics* 12 (1996): 510–15.

p. 233 Genetic engineering has been attacked: My responses to these objections are the same as my responses to similar attacks on embryo selection and will not be repeated here.

p. 234 . . . it is wrong to tread in God's domain: *The Far Side* cartoon created by Gary Larson entitled "Young Victor Frankenstein stays after school" shows Victor at the blackboard in the front of the classroom writing, "I will not play in God's domain; I will not play in God's domain; . . . I will not play in God's dom—" (syndicated on January 27, 1996).

p. 234 . . . "Mother Nature," an updated feminine personification of God himself: Actually, the view that Nature and God are one and the same has been propounded since classical times and is referred to as "pantheism."

p. 234 . . . Christian Scientists . . . reject of all forms of medical treatment for disease: Christian Scientists do make an exception when it comes to teeth and eyes, or the setting of bones. Essentially what they reject is the use of medicines—in the form of chemicals—to cure disease.

p. 234 By 1994, fully 75 percent of all Americans: Results of a survey conducted for *Family Circle* magazine by Princeton Survey Research Associates in May 1994.

p. 234 the injection of testis-derived nuclei into the egg cytoplasm: This ancillary technology is used in those cases where a man is unable to produce sperm.

p. 235 45 percent of Americans reject all uses of genetic engineering: However, another survey conduced by the March of Dimes Foundation in 1992 found that 87 percent of those polled understood "little or nothing about gene therapy." This result suggests that some respondents to questions about genetic engineering may not understand what it is that they are being asked.

p. 235 The British public is even more wary: Theresa Marteau, Susan Michie, Harriet Drake, and Martin Bobrow, "Public Attitudes Towards the Selection of Desirable Characteristics in Children," *Journal of Medical Genetics,* 32 (1995): 796–98.

p. 235 *"Just as the Christian soul has provided an archetypal concept"*: Dorothy Nelkin and M. Susan Lindee, *The DNA Mystique: The Gene as a Cultural Icon* (New York: W. H. Freeman, 1995), pp. 41–42 (paperback version).

p. 237 The number of parents: Remember that in most cases, embryo selection, rather than genetic engineering, will be sufficient to prevent the appearance of lethal diseases.

p. 237 "The human species has a deeply ingrained tendency to prove the experts wrong": Freeman Dyson, *Infinite in all Directions* (New York: Harper & Row, 1988), p. 126.

p. 238 . . . *radiotelepathy,* the term used and defined by Freeman Dyson: Ibid, pp. 130–37.

EPILOGUE: HUMAN DESTINY?

p. 241 "the gradual widening of the present merely temporary and social difference": Original reference is H. G. Wells (1895) *The Time Machine.* Quote can be found on p. 301 of *Three Prophetic Science Fiction Novels of H. G. Wells* (New York: Dover, 1960).

p. 246 . . . the successful incorporation of photosynthetic units into embryos: The photosynthetic units of plant cells are contained within little organelles called chloroplasts that float in the cytoplasm. All chloroplasts can trace their ancestry back to an single-cell photosynthetic creature that was gulped into a larger nonphotosynthetic cell. But instead of the smaller cell being eaten by the larger cell, the two set up a symbiotic relationship with each other that was so successful, it served as the starting point for the entire plant kingdom. Remarkably, as a lasting remnant of their independent origin, chloroplasts still retain their very own little genomes.

p. 247 . . . far-flung communities will begin to lose contact with one another: Freeman Dyson explains why: "Even messages traveling at the speed of light take fifty thousand years to creep across the galaxy. Whole historical epochs will pass, cultures will rise and fall, between a telephone call and the reply. Each little piece of the galaxy will be a world of its own, isolated from other pieces by the immensity of space and the quickness of time. We shall enjoy abundant communication with our neighbors in the past, but of our neighbors in the present we can know nothing." Freeman Dyson, *Imagined Worlds* (Cambridge, Massachusetts: Harvard University Press, 1997), p. 163.

Index

About the Author

Lee M. Silver is a professor at Princeton University in the Departments of Molecular Biology, Ecology and Evolutionary Biology, and the Program in Neuroscience. He received a bachelor's degree and master's degree in physics from the University of Pennsylvania and a doctorate in biophysics from Harvard University. He conducts research in genetics, evolution, reproduction, and developmental biology, with a current focus on behavioral genetics. Professor Silver teaches courses and lectures widely on the social impact of biotechnology, with an emphasis on reprogenetics. In 1993, he was elected a fellow of the American Association for the Advancement of Science. He lives with his wife and three children in Princeton, New Jersey.